MICROMECHANICS OF SOLIDS

MICROMECHANICS OF SOLIDS

D. R. AXELRAD

Thomas Workman Professor
of Mechanical Engineering
McGill University

ELSEVIER SCIENTIFIC PUBLISHING COMPANY
Amsterdam–Oxford–New York
PWN—POLISH SCIENTIFIC PUBLISHERS
Warszawa
1978

Graphic design: Zygmunt Ziemka

Distribution of this book is being handled by the following publishers:

for the U.S.A. and Canada
ELSEVIER/NORTH-HOLLAND, INC.
52 Vanderbilt Avenue
New York, New York 10017

for Albania, Bulgaria, Chinese People's Republic, Czechoslovakia, Cuba, German Democratic Republic, Hungary, Korean People's Democratic Republic, Mongolia, Poland, Rumania, U.S.S.R., Vietnam and Yugoslavia
ARS POLONA,
Krakowskie Przedmieście 7, 00-068 Warszawa, Poland

for all remaining areas
ELSEVIER SCIENTIFIC PUBLISHING COMPANY
335 Jan van Galenstraat
P.O. Box 211, Amsterdam, The Netherlands

Library of Congress Cataloging in Publication Data

Axelrad, David Robert
Micromechanics of solids.

Bibliography: p.
Includes index.
1. Micromechanics. 2. Solids. I. Title.
QC176.8.M5A97 530.1'2 77-22590
ISBN 0-444-99806-3

Printed in Poland by D.R.P.

To Kay and Jean
For their patience and understanding during the writing of this text.

Preface

This monograph grew out of lectures, which the author gave for a number of years to graduate students in Micromechanics at McGill University and various Institutes in the U.S.A., Canada and Europe. Its purpose is to present the probabilistic micromechanics of structured media in a formal manner and to provide a summary of the theoretical and experimental research that has been carried out over the years in the Micromechanics of Solids Laboratory since the introduction of the micromechanics theory by the author.

The author originally intended to include a somewhat larger exposition of experimental micromechanics. However, due to the limitation in the desired size of this volume, it was decided that the concluding chapter of the text concerned with the experimental investigations should be kept to a minimum.

For the sake of brevity the emphasis in the presentation is on general results rather than on detailed examples. For the reader who has had no contact with the mathematical theory of probability and the theory of stochastic processes, a brief review of the important concepts of these theories required for the subsequent development of probabilistic micromechanics, is made available in the first chapter of the text. The actual random theory of deformation and the underlying concepts are dealt with in Chapter II. Due to the development of a more general deformation theory of structured media, it has been necessary to introduce in Chapter III some topological considerations and certain concepts of functional analysis. Chapter IV is concerned with some applications of the micromechanics theory including the mechanical relaxation of structured solids. In Chapter V an attempt is made to validate various hypotheses and concepts of the micromechanics theory on the basis of experimental evidence. The latter was obtained by using various experimental techniques which for the reason given above will only be briefly discussed.

The author acknowledges the National Research Council of Canada and the Faculty of Graduate Studies and Research of McGill University, without whose generous support the founding of the Micromechanics of Solids Laboratory would not have been possible.

It is also a pleasure to acknowledge Prof. H. Zorski, of the Polish Academy of Sciences, Prof. W. Olszak, of the International Centre for Mechanical Sciences, Udine, and Prof. Th. Lehmann, of the Institute of Mechanics, Ruhr-University, Bochum, for their kind invitations at various times to deliver seminars and courses at their Institutes.

The author would like to express his sincere thanks to his colleagues Dr. J. W. Provan and Dr. D. Atack, Director of the Applied Physics Division, Pulp and Paper Research Institute of Canada, for many valuable discussions and contributions to this work. He would also like to extend his thanks to past and present graduate students and research associates, Drs. J. Kalousek, S. Basu and Y. Haddad. Special thanks are due to Dr. S. Basu for proof reading the entire manuscript and for his conscientious assistance throughout its preparation. Mr. Ri. Peralta-Fabi kindly supplied all diagrams presented in this text for which the author would also like to thank him.

Special acknowledgement is made to the publishers, Polish Scientific Publishers, (PWN), Warsaw and Elsevier Publ. Co., Amsterdam, for their support and in particular to their editor, Prof. H. Zorski for his valuable suggestions.

Finally, the author expresses his thanks to Mrs. M. L. Powell for so patiently typing the original manuscript.

D. R. AXELRAD

Contents

"One must start from the concept; and even if it cannot possibly do justice to the 'rich variety' of nature, as the saying goes, one must trust the concept, though much that is particular cannot yet be explained ... The concept is valid by itself; the particulars will then surely fall in line."

G. W. F. Hege
(1770–1831)

Encyclopedia of the Philosophical Sciences (par. 353)

I. Introduction to the Probabilistic Micromechanics of Solids

1.1 Introduction

One of the main objectives of material science is the formulation of stress deformation relations that govern the mechanical response of a given material under specific environmental conditions. Traditionally in the mechanics of solids, models for the prediction of the response behaviour have been used, that are based on continuum theory. In general, these refer to homogeneous media, ignoring thereby the presence of the microstructure of the material.

Probabilistic micromechanics is equally concerned with the formulation of the response of various classes of solid media but with the inclusion of microstructural effects that are due to the inherent geometrical and physical properties of structured solids. Most of the significant field quantities involved in any formulation of the material behaviour are by nature random variables or functions of such variables.

Hence the approach formally presented in this monograph employs from the onset probabilistic concepts and principles of statistical mechanics.

In the past, several attempts have been made to modify the classical continuum approach by allowing for microscopic or "local" quantities to enter into the analysis, but without removing the main restrictions imposed by continuum physics on such formulations. In contrast to these "modified continuum theories", other researchers have attempted to account for the randomness of the microstructure and to arrive, on the basis of classical statistical mechanics, to a formulation that reflects to a certain extent the macroscopic response of structured media.

In probabilistic micromechanics, structured media are considered as discrete systems whereby the occurring interaction effects between structural elements, defined more specifically later in this text, are taken into account. Such an approach is more appropriate to real materials and is in line with experimental observations clearly indicating the internal discontinuities such materials always have. As mentioned previously, since the characteristic quantities are considered as random variables or random functions, the deformation process itself is seen as a stochastic process. In particular, within the range of the completely reversible or irreversible steady-state deformations a material may undergo, the deformation process is regarded as a Markov process. However within the intermediary range, e.g. between these two stages of deformation, the material response can only be treated in terms of a discontinuous random process in which a limit analysis and the concept of transition functions become significant.

Since the formulation as presented in this text begins with the analysis of the response of individual "microelements" that form a given microstructure of the medium and since experimental observations are usually evaluated in terms of macroscopic quantities, a connection between the two descriptions becomes necessary. For this purpose the notion of "intermediary domains" referred to as mesodomains is introduced. A denumerable number of such non-intersecting domains form then the macroscopic material body. This concept and the postulate of their existence within the medium form the link between the microscopic and macroscopic description of the material behaviour. A more detailed discussion of this and other concepts that are fundamental in the probabilistic micromechanics of solids is given later in Chapter II of the monograph. From a thermodynamics point of view it is equally important to find a connection between the microscopic and macroscopic formulation. In contrast to the classical formulation, in the present study the analysis is based on "statistical micromechanics", which, analogous to the classical representation, aims at the formulation of a set of "Governing equations for structured solids". This will be seen possible by assuming the Markovian character of the deformation process.

It remains to remark briefly on the analytical methods employed in probabilistic micromechanics. Considerable use is made of the theorems of the mathematical theory of probability and the principles of statistical mechanics. It has been found useful in the formulation of the response behaviour of structured solids to employ the operational representation of various relations, which in turn requires functional analysis and topological considerations. In order to clarify such operational relations some of the fundamental aspects of functional analysis are given in Chapter III of the text. The applications of probabilistic micromechanics for two classes of structured solids only are given in Chapter IV. The choice of these two groups of materials is mainly due to the considerable interest in their response behaviour for industrial applications. These two classes of materials will also be investigated with respect to their mechanical relaxation behaviour.

Finally, it should be noted that throughout the text direct notation will be employed in which bold type letters indicate vectors or tensors as the case may be, but whenever necessary index notation will also be introduced. Since the main purpose of this monograph is to provide a compact summary of the theory, detailed derivations of various relations have been kept to a minimum in order to highlight the basic principles involved.

1.2 Classes of materials in micromechanics of solids

Any solid material that exhibits a heterogeneous structure can be considered, so far as the response characteristics are concerned, from the point of view of probabilistic micromechanics. With some idealization of the complex microstructure such materials may have, the response behaviour can be described in terms of the theory discussed in this text. In order to narrow down the great variety of structured solids, considerations are given here to an approximate classification as shown in Table I below. In particular, at present the research efforts in the Micromechanics Laboratory are concentrated on Groups I and III. As mentioned previously the theoretical and experimental approaches discussed subsequently will be mainly concerned with these classes of materials.

TABLE I

Some classes of structured solids

I	II	III	IV	V	VI
Polycrystalline solids	Composite materials	Fibrous systems	Polymeric materials	Particulate materials Soils	Bio-materials
High temperature solids	Whiskers and particles embedded in suitable matrix	Paper	Synthetic fibrous structures		Bone structures, tissues
Directional solidified metals		Textiles		Dispersed particle systems	Muscle fibres

Since its inception by the author of this monograph (1966) probabilistic micromechanics has been chiefly concerned with polycrystalline solids. Such materials under moderate magnification after etching and polishing appear as shown in Fig. 1 representing a Silicon steel

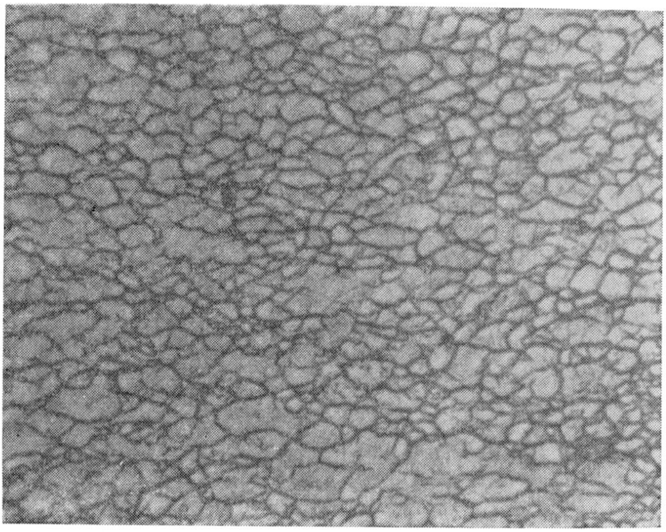

Fig. 1. Silicon steel at approx. 800°C. Micrograph (magnification 585×).

at an elevated temperature (magnification: 585×). It reveals a clearly defined substructure similar to that at lower temperatures and it can be immediately seen that two structural regions exist, e.g. the basic structural unit or "microelement" and the boundary surrounding each microelement. A more distinct view of an arbitrary area of the micrograph in Fig. 1 is shown in Fig. 2 below, obtained by Trans-

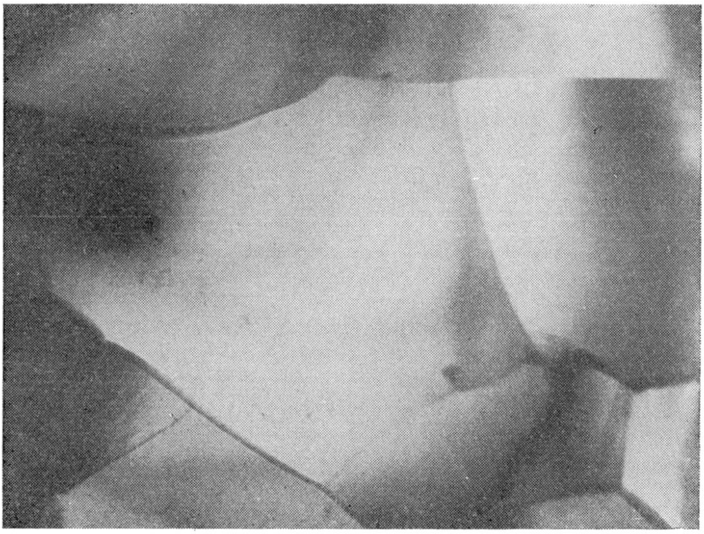

Fig. 2. Silicon steel at approx. 800°C. Micrograph of an arbitrary small area (Fig. 1) under Transmission Electron Microscope (magnification 5700×).

mission Electron Microscopy (TEM) at a magnification of 5700×. Hence in such a material a structural element is considered as a "microelement" which is surrounded by the "grain boundary".

From the point of view of probabilistic micromechanics both these regions must be taken into account in the formulation of the mechanical response of such materials. It is well known from material science that the crystallographic orientation and the presence of defects such as dislocations in each crystal play a fundamental role in the mechanical response characteristics. Such effects should be incorporated therefore in any theory of deformation that aims at the inclusion of the microstructural characteristics. Furthermore, the fact that the "relative" crystallographic orientation of two adjacent crystals or microelements is of significance in the response mechanism of the grain boundary, considerations of this orientation should also be included in the analysis of the deformation process.

For the analytical study of polycrystalline solids as treated in subsequent chapters, a model will be used as indicated in Fig. 3 below.

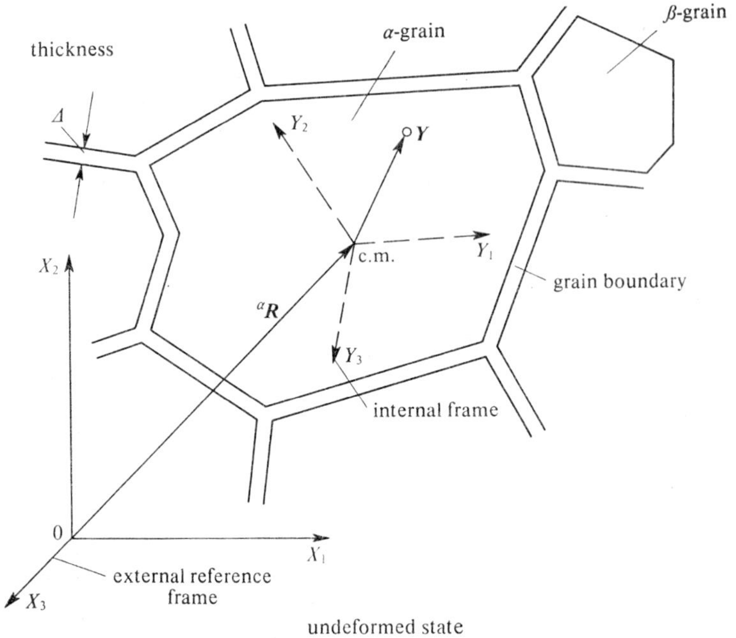

Fig. 3. Single grain or "microelement" in the undeformed state.

This model shows the individual grain "α" and adjacent ones, which form an ensemble in the intermediary domain or "mesodomain" of the material body. It also clearly indicates a delineation of the grains by grain boundaries of a certain thickness "Δ" in the underformed state of the material. It is convenient in the subsequent deformation kinematics of such solids to use an internal reference frame (Y_1, Y_2, Y_3) located at the centre of mass of a grain "α" and an external frame (X_1, X_2, X_3) with respect to which the deformations can be formulated.

More recently, attention has turned to fibrous systems (Group III of Table I) and, in particular, to the microstructural effects in the overall response characteristics of sheet materials of such solids including the behaviour of single fibres and bonding areas. Again a typical micrograph of such a network as obtained from a Scanning Electron Microscope with a magnification of 170× is shown in Fig. 4. It represents a strip of beaten sulphite paper, bleached and

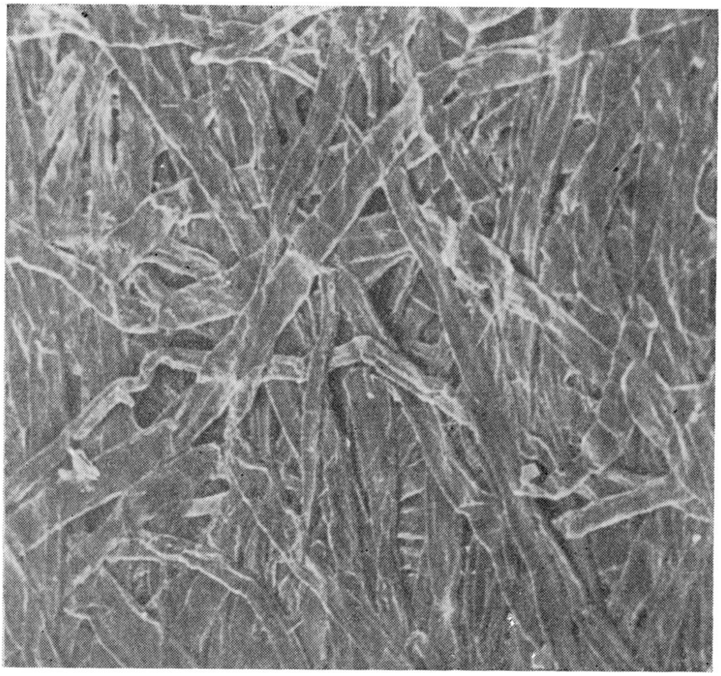

Fig. 4. Fibrous system Scanning Electron Microscope photograph (magnification 170×).

dried as frequently occurring in the paper industry. The subsequent analysis of such systems is limited to a fibrous network that has strong bonding in the plane of the sheet (Fig. 5) but a much weaker one perpendicular to it. This is the case of a fibrous system such as paper. Fig. 5 below also indicates the choice of a structural or microelement, which in this case is considered to consist of a certain fibre segment "α" and two halves of the effective bond area between overlapping fibres α, β. It is here also convenient for the study of the deformation kinematics of such systems to attach a body frame Y_1, Y_2 at the centre of mass of the bond area and an external frame X_1, X_2 as shown schematically in Fig. 5 for the undeformed state of the

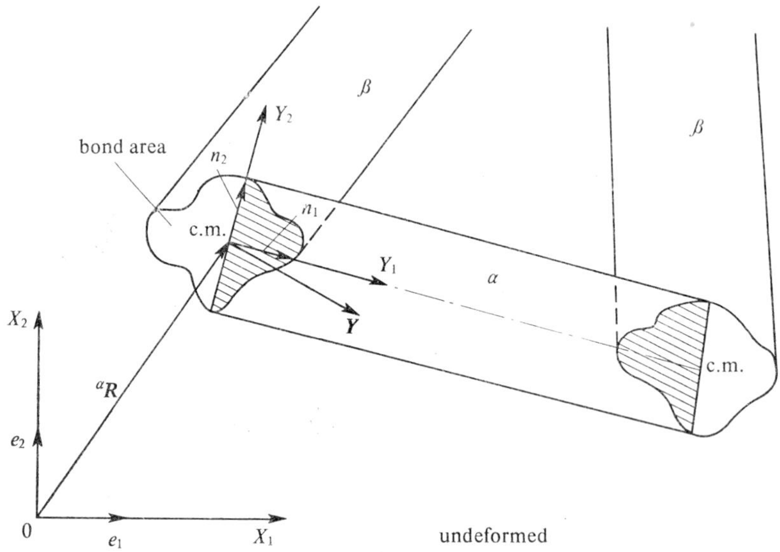

Fig. 5. Microelement of a fibrous system (2-dimensional network).

network. The micrograph shown in Fig. 6 represents an actual bonding area between two fibres for an unbeaten, low-yield sulphite SF-paper to a magnification of approximately 1120× in the Scanning Electron Microscope. It reveals further the necessity of considering the bonding effect in the deformation analysis, since this effect contributes considerably to the overall mechanical strength of the fibrous network.

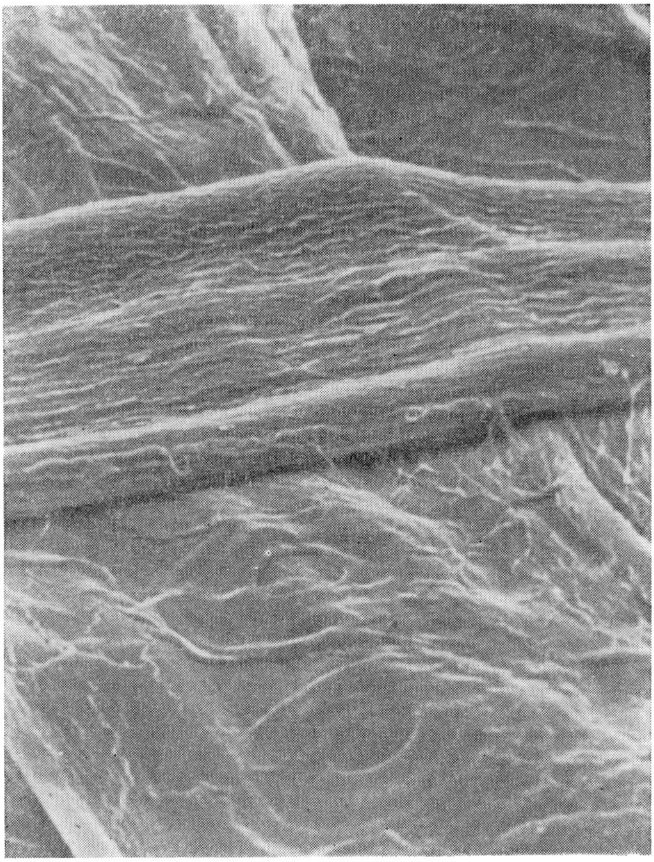

Fig. 6. Micrograph of actual bonding area by Scanning Electron Microscopy (unbeaten low-yield Sulphite SF-paper, magnification 1120×).

1.3 Probabilistic concepts in micromechanics

The characteristic field variables involved in the analysis of the response behaviour of structured solids are either scalars, vectors or tensor valued quantities. As mentioned earlier (Section 1.1) in probabilistic micromechanics these quantities are considered as random variables or functions of such variables. It appears appropriate therefore to briefly review some of the important concepts of the mathematical theory of probability and the theory of stochastic processes.

Such a review, whilst kept to a minimum, is also required for the introduction of certain notations and definitions employed later in the analysis of structured solids. For a more comprehensive study the reader is referred to the texts listed in the bibliography of this monograph[1-10].

(A) *Notations and definitions*

(i) *Trial*

The performance of a random experiment is referred to as a trial. Such an experiment is carried out without any *a priori* knowledge. The result of such an experiment is called an "outcome ω".

(ii) *Sample space*

A sample space is a set of all possible outcomes ω denoted here by $\Omega = \{\omega\}$.

(iii) *Event*

An event is a subset of Ω denoted by $E \subset \Omega$ such that $E = \{\omega \in \Omega;$ ω satisfies certain properties$\}$. Thus making the statement that an event E occurs means that the outcome ω of the random experiment is an element of E. Events may be simple or indecomposable or they may be compound or decomposable. A simple event is one with only one outcome.

(iv) *Empty set*

The empty set is denoted by $\emptyset$ and since it contains no outcome ω, it will be called an "impossible event". The empty set is added to the sample space and for the sake of simplicity the same notation Ω is used for the extended space.

Definitions (i)–(iv) introduce set-theoretical concepts, typical of the theory of probability. For the reader's convenience we present now the fundamental set-theoretical operations over the events.

(v) *Complement set (event)*

The complement of an event E is another event $\bar{E}$ such that $\bar{E} = \{\omega \in \Omega; \omega \notin E\}$. It should be noted that: $\bar{\Omega} = \emptyset$ and $\bar{\emptyset} = \Omega$.

(vi) *Set union*

The union of two sets E_1 and E_2 is again a set E_3 which contains both the elements of E_1 and E_2 inclusive, e.g. $E_3 = \{\omega \in \Omega; \omega \in E_1 \cup \cup\, \omega \in E_2\}$ or symbolically written as $E_3 = E_1 \cup E_2$.

(vii) *Set intersection*

The intersection of two sets E_1 and E_2 is a set E_3 which contains only those elements that are common in E_1 and E_2, expressed by:

$$E_3 = \{\omega \in \Omega; \omega \in E_1 \cap \omega \in E_2\}$$

again symbolically written as:

$$E_3 = E_1 \cap E_2$$

It should be noted that the symbol $\subset$ means a set inclusion, whilst $\in$ means an element inclusion. On basis of the concepts of "mutually exclusive sets" and set inclusion one can establish the "equality of sets" as shown below.

(viii) *Set inclusion*

Two sets E_1 and E_2 are inclusive, if $E_1 \cap E_2 \neq \emptyset$ or if $E_1 \cap E_2 = E_3 = \{\omega \notin \emptyset\}$.

(ix) *Mutually exclusive sets*

Two sets E_1 and E_2 are mutually exclusive, if

$$E_1 \cap E_2 = E_3 = \emptyset$$

(x) *Set equality*

Two sets E_1 and E_2 are equal, if every element of E_1 is also an element of E_2 and every element of E_2 is an element of E_1, i.e. $E_1 = E_2$ implies $E_1 \subset E_2$, $E_2 \subset E_1$. Symbolically this is expressed as $\omega \in E_1 \Leftrightarrow \omega \in E_2$ for $\forall \omega \in \Omega$.

(B) *Probability concepts*

There are several inductive ways to define "probability". However in the mathematical theory of probability two distinct definitions are used. The first due to v. Mises[1] is based on the frequency interpretation, whilst the second is derived more strictly from the point of view of measure theory and is an axiomatic definition due to Kolmogorov[2].

(i) *Frequency definition*

If a random experiment is carried out n times under identical conditions and if the event E_i occurs n_i times, then the probability of the event E_i written as $\mathscr{P}\{E_i\}$ is given by:

$$\mathscr{P}\{E_i\} = \lim_{n\to\infty} \frac{n_i}{n} \tag{1.1}$$

Hence this concept of the probability of an event E_i to occur under fixed experimental conditions is the relative frequency n_i about which the real frequency of this event becomes stable as the number of trials $n \to \infty$. Whilst n may be large it is never infinite. Hence a more strict definition is obtained by postulating that above form holds almost everywhere.

(ii) *Axiomatic definition*

$$\mathscr{P}\{E_i\} \overset{\text{a.e.}}{=} \lim_{n\to\infty} \frac{n_i}{n} \tag{1.2}$$

which is based on the following three postulates (see Kolmogorov[2]):

I. $\mathscr{P}\{E_i\} \geqslant 0$ (0 being an impossible event)

II. $\mathscr{P}\{\Omega\} = 1$ (1 being always a certain event)

III. For two mutually exclusive events E_i, E_j

$$\mathscr{P}\{E_i \cup E_j\} = \mathscr{P}\{E_i\} + \mathscr{P}\{E_j\}, \quad E_i \cap E_j = \emptyset, \quad i \neq j$$

The second definition combined with the laws of measure theory is a more rigorous definition used in the mathematical theory of probability. Definitions (i) and (ii) are the consecutive stages of a process of abstraction beginning with the concept of the frequency n_i/n of an event and ending with the introduction of the concept of the probability space.

(C) *Probability space and measure*

It may be noted from the foregoing remarks that although $E \subset \Omega$ every subset of Ω is not necessarily an event. In order to study this restriction it will be necessary to consider the concepts of probability space and probability measure. In this context the definition of a "σ-field" is required first. Thus consider an event E_1 such that $\bar{E}_1$ is also an event. If E_2 is another event, then $E_1 \cup E_2$ and $E_1 \cap E_2$ will also be an event. Hence a non-empty class of sets that satisfies these conditions is termed a field $\mathscr{F}$, defined by:

$$\mathscr{F} = \{E_i; i = 1, 2, \ldots\}$$

such that

$$E_i \in \mathscr{F} \Rightarrow \bigcup_{i=1}^{\infty} E_i \in \mathscr{F}$$

More specifically a:

(i) *σ-field* (*σ-ring*)

is a non-empty class $\mathscr{F}$ of sets E_i such that:

I. $E_i \in \mathscr{F} \Rightarrow \bar{E}_i \in \mathscr{F} \quad i = 1, 2, \ldots$

II. $E_i \in \mathscr{F}$ and $E_j \in \mathscr{F} \Rightarrow E_i \cap E_j \in \mathscr{F}$

III. $E_i \in \mathscr{F} \Rightarrow \bigcup_{i=1}^{\infty} E_i \in \mathscr{F}$

Hence a σ-field $\mathscr{F}$ is closed under the operations of countable intersection, union and complement.

(ii) Using the above postulates of a σ-field a somewhat stronger axiomatic definition of probability can now be given as follows:

I. $\mathscr{P}\{E_i\} \geqslant 0, \quad \mathscr{P}\{E_i\} = 0 \Leftrightarrow E_i = \emptyset$

II. $\mathscr{P}\{\Omega\} = 1$

III. $\mathscr{P}\{\bigcup_{i=1}^{\infty} E_i\} = \sum_{i=1}^{\infty} \mathscr{P}\{E_i\}, \quad \text{if} \quad E_i \cap E_j = \emptyset, \quad i \neq j$

(iii) *Measurable space*

The sample space Ω that contains all possible outcomes together with the above defined σ-field $\mathscr{F}$ of events E is called a measurable space denoted by $(\Omega, \mathscr{F})$.

(iv) *Probability space*

A probability space or equivalently a "measure space" is a measurable space $(\Omega, \mathscr{F})$ as defined above together with an appropriate probability measure in accordance with its axiomatic definition (ii) and is denoted by the triplet $(\Omega, \mathscr{F}, \mathscr{P})$. It should be noted that a probability measure is a "set function", i.e. it is defined for events and not for outcomes. In this context some useful relations may be given as follows:

(a) $\mathscr{P}\{\emptyset\} = 0$

(b) $\mathscr{P}\{\bar{E}_i\} = \mathscr{P}\{\Omega\} - \mathscr{P}\{E_i\}$

(c) $\mathscr{P}\{E_i \cap E_j\} = \mathscr{P}\{E_i\} - \mathscr{P}\{E_i \cap \bar{E}_j\}$

(d) $\mathscr{P}\{E_i \cup E_j\} = \mathscr{P}\{E_i\} + \mathscr{P}\{E_j\} - \mathscr{P}\{E_i \cap E_j\}, \quad \text{if} \quad E_i \cap E_j \neq \emptyset$

(e) $\mathscr{P}\{E_i\} \leqslant \mathscr{P}\{E_j\} \quad \text{if} \quad E_i \leqslant E_j$

(f) $\sum_{i=1}^{N} \mathscr{P}\{E_i\} = \mathscr{P}\{\bigcup_{i=1}^{N} E_i\} = 1, \quad \text{if} \quad \sum_{i=1}^{N} E_i = \Omega$

and $E_i \cap E_j = \emptyset, \quad i \neq j, \quad i = 1, \ldots, N, \quad j = 1, \ldots, N$

A more detailed discussion on the probability measure, and the choice of an appropriate probability space as a function space will be given in Chapter III, in conjunction with the random theory of deformation and the application of functional analysis in probabilistic micromechanics in general.

(D) *Conditional probability*

The notion that a "conditional probability" should be taken as the fundamental quantity in the mathematical theory of probability was first introduced by Rényi[5]. This notion originated from the fact that the probability of an event depends largely on the circumstances under which the occurrence or non-occurrence of such an event is observed. Hence the axiomatic theory of "conditional probability spaces" is introduced, which are generated by "bounded measures". The previously introduced probability space and corresponding measure are thus seen as a special case of this more general theory of probability. For a more detailed study the reader is referred to references[5,11,12]. In line with the intended brief review of probabilistic concepts in micromechanics the following definitions will be considered.

(i) *Joint probability*

Using the frequency interpretation of probability (1.1), consider now two events E_1 and E_2. Let N_1 denote the number of outcomes favourable to E_1 in an experiment and N_2 the number of outcomes favourable to the event E_2. Then

$$\mathscr{P}\{E_1\} = \frac{N_1}{N_\Omega}$$
$$\mathscr{P}\{E_2\} = \frac{N_2}{N_\Omega} \tag{1.3}$$

in which N_Ω is the number of total outcomes. If E_1 and E_2 are not mutually exclusive, then $E_1 \cap E_2 \neq \emptyset$. In that case if N_{12} denotes the number of outcome that is favourable to both E_1 and E_2, the joint probability can be defined by:

$$\mathscr{P}\{E_1 \cap E_2\} = \frac{N_{12}}{N_\Omega} \tag{1.4}$$

In view of the above definitions, $\mathscr{P}\{E_1\}$ and $\mathscr{P}\{E_2\}$ are referred to as "marginal probabilities" of E_1 and E_2, respectively.

(ii) *Conditional probability*

The conditional probability of an event E_1 with respect to an event E_2 denoted by $\mathscr{P}\{E_1|E_2\}$ represents the probability that event E_1 occurs under the condition that E_2 has already occurred. Hence in this case only the outcomes favourable to E_2 are considered and not all outcomes. This can be expressed using again the frequency interpretation of probability as follows:

$$\mathscr{P}\{E_1|E_2\} = \frac{N_{12}}{N_2} = \frac{N_{12}/N_\Omega}{N_2/N_\Omega} = \frac{\mathscr{P}\{E_1 \cap E_2\}}{\mathscr{P}\{E_2\}} \tag{1.5}$$

Similarly

$$\mathscr{P}\{E_2|E_1\} = \frac{N_{12}}{N_1} = \frac{N_{12}/N_\Omega}{N_1/N_\Omega} = \frac{\mathscr{P}\{E_1 \cap E_2\}}{\mathscr{P}\{E_1\}} \tag{1.6}$$

It can be easily shown from the above relations that:

$$\mathscr{P}\{E_1|E_2\} = \frac{\mathscr{P}\{E_2|E_1\}\mathscr{P}\{E_1\}}{\mathscr{P}\{E_2\}} \tag{1.7}$$

The above expression is one form of the well-known Bayes rule. The above formulas can also be obtained by using the axiomatic definitions discussed earlier.

(iii) *Independent events*

Considering N mutually exclusive sets E_i, $i = 1, \ldots, N$ in Ω such that

$$\sum_{i=1}^{N} E_i = \Omega \tag{1.8}$$

Then if E is an arbitrary event in Ω, it follows that:

$$\mathscr{P}\{E\} = \sum_{i=1}^{N} \mathscr{P}\{E|E_i\}\mathscr{P}\{E_i\} \tag{1.9}$$

Now, two events are said to be "pairwise independent", if

$$\mathscr{P}\{E_i \cap E_j\} = \mathscr{P}\{E_i\}\mathscr{P}\{E_j\}, \quad i \neq j \tag{1.10}$$

However, it does not follow strictly that pairwise independence implies "absolute independence". The latter case is characterized by:

$$\mathscr{P}\{E_{k_1} \cap E_{k_2} \cap \ldots \cap E_{k_n}\} = \mathscr{P}\{E_{k_1}\}\mathscr{P}\{E_{k_2}\} \ldots \mathscr{P}\{E_{k_n}\} \tag{1.11}$$

1.4 Random variables and functions of random variables

(A) *Random variables and random vectors*

A random variable (r.v.) x is a mapping that assigns to each outcome $\omega \in \Omega$ of a random experiment a real number $x(\omega) \in R$. The real number $x(\omega)$ is the value of the r.v. x. The mapping x is required to be such that for numbers $\alpha \in R$ the sets

$$E_\alpha \equiv \{x(\omega) \leqslant \alpha\} = \{\omega \in \Omega; x(\omega) \leqslant \alpha\}$$

have the properties

$$\begin{aligned} &\text{(i) } E_\alpha \text{ is an event in } \mathscr{F} \text{ for every } \alpha \in R \\ &\text{(ii) } \lim_{\alpha\to-\infty} \mathscr{P}\{E_\alpha\} = 0, \quad \lim_{\alpha\to\infty} \mathscr{P}\{E_\alpha\} = 1 \end{aligned} \tag{1.12}$$

A random vector $\boldsymbol{x}$ associated with $(\Omega, \mathscr{F}, \mathscr{P})$ is a real-valued vector function that maps Ω into R^n such that the sets

$$E_{\boldsymbol{\alpha}} \equiv \{\boldsymbol{x}(\omega) \leqslant \boldsymbol{\alpha}\} = \{\omega \in \Omega; \boldsymbol{x}(\omega) \leqslant \boldsymbol{\alpha}\}$$

have the properties

$$\begin{aligned} &\text{(i) } E_{\boldsymbol{\alpha}} \text{ is an event in } \mathscr{F} \text{ for every } \boldsymbol{\alpha} \in R^n \\ &\text{(ii) } \lim_{\boldsymbol{\alpha}\to-\infty} \mathscr{P}\{E_{\boldsymbol{\alpha}}\} = 0, \quad \lim_{\boldsymbol{\alpha}\to\infty} \mathscr{P}\{E_{\boldsymbol{\alpha}}\} = 1 \end{aligned} \tag{1.13}$$

where for vectors $\boldsymbol{a} = (a_1, \ldots, a_n)$ and $\boldsymbol{b} = (b_1, \ldots, b_n) \subset R^n$ the inequality $\boldsymbol{a} \leqslant \boldsymbol{b}$ means that $a_i \leqslant b_i$ for $i = 1, \ldots, n$ and $\boldsymbol{\alpha} = (\alpha_1, \ldots, \alpha_n) \to \pm\infty$ means that $\alpha_i \to \pm\infty$ for $i = 1, \ldots, n$.

(B) *Distribution and density functions*

(i) *Distribution functions*

The distribution function $F_x(\alpha)$ of a random variable x is defined by:

$$F_x(\alpha) = \mathscr{P}\{x \leqslant \alpha\} \quad \text{for all } \alpha \in R \tag{1.14}$$

Hence, it is the probability that the value $x(\omega)$ of the random variable x is less than or equal to α, where α is an element of R.

The distribution function $F_x(\alpha)$ has the following properties:

I. $F_x(-\infty) \equiv \lim_{\alpha\to-\infty} F_x(\alpha) = 0$

II. $F_x(\infty) \equiv \lim_{\alpha\to\infty} F_x(\alpha) = 1$

III. $F_x(\alpha)$ is a non-decreasing function

If we consider properties of a definite random variable or when we want to simplify the notation, provided it cannot lead to any confusion, we shall replace the symbol $F_x(\alpha)$ by $F(\alpha)$, i.e.

$$F_x(\alpha) \equiv F(\alpha) \tag{1.15}$$

A distribution function can be either discrete or continuous as indicated in Fig. 7 (a, b) below.

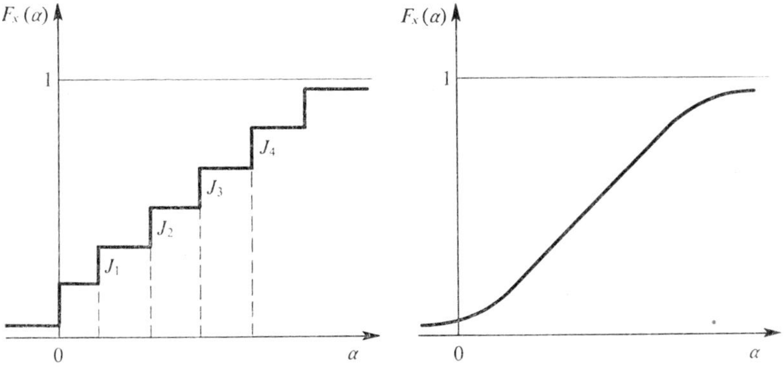

(a) Discrete distribution (b) Continuous distribution

Fig. 7. Types of distribution functions.

In accordance with the properties of distribution functions (I–III), a continuous random variable x can be defined as one for which $F_x(\alpha)$ is continuous for all $\alpha \in R$ and where derivatives of $F_x(\alpha)$ exist, except at a countable number of points. Thus $F_x(\alpha)$ is continuous from the right, i.e.

$$\lim_{\alpha \downarrow \alpha_0} F_x(\alpha) = F_x(\alpha_0^+) = F_x(\alpha_0) \tag{1.16}$$

A discrete random variable is defined as one for which $F_x(\alpha)$ is discrete for $\forall \alpha \in R$ and where derivatives of $F_x(\alpha)$ exist and are equal to zero, except at a countable number of "jump points" (Fig. 7(a)). It may be noted that most of the observable variables in experimental micromechanics have discrete distribution functions.

(ii) *Density functions*

The density function of a random variable x is denoted by $p_x(\gamma)$ and is defined as follows:

$$F_x(\alpha) = \sum_{\gamma_i \leqslant \alpha} p_x(\gamma_i) \quad \text{for the discrete case of } F_x(\alpha) \text{ and for every } \alpha \in R,\ i = 1, 2, \ldots, N = N(\alpha)$$

$$F_x(\alpha) = \int_{-\infty}^{\alpha} p_x(\gamma)\,\mathrm{d}\gamma \quad \text{for the continuous case of } F_x(\alpha) \text{ and for every } \alpha \in R,\ \gamma \in (-\infty, \alpha) \tag{1.17}$$

In the discrete case the function $F_x(\alpha)$ is constant in the neighbourhood of every point $\alpha \in R$ except at most a denumerable set of points $\alpha = \gamma_i$ where its value has a jump. Alternatively one can write the density functions

$$p_x(\gamma_i) = F_x(\gamma_i^+) - F_x(\gamma_i^-) \quad \text{in the discrete case}$$

$$p_x(\alpha) = \frac{\partial F_x(\alpha)}{\partial \alpha} \quad \text{in the continuous case} \tag{1.18}$$

where $F_x(\gamma_i^{\pm})$ denotes the right-hand (+) and the left-hand (−) limits of $F_x(\alpha)$ at the point $\alpha = \gamma_i$, respectively. The notations used in the above definitions can be simplified by writing for the discrete distribution and density functions: F_i or $F(\alpha_i)$ and p_i or $p(\alpha_i)$, respectively. Similarly for the continuous case one may write $p_x(\alpha)$ or $p(\alpha)$. The density functions have the following properties:

$$\text{I. } p_i \text{ or } p(\alpha) \geqslant 0$$

$$\text{II. } \sum_{i=1}^{N} p_i = 1,\ N \leqslant \infty \text{ for the discrete case} \tag{1.19}$$

$$\text{III. } \int_{-\infty}^{+\infty} p(\alpha)\,\mathrm{d}\alpha = 1 \text{ for the continuous case}$$

(iii) *Joint distribution and density functions*

The "joint distribution and density functions" arise if, for example, one considers the multi-dimensional case of a random variable. Thus in a similar manner as above the distribution function of a random vector $\boldsymbol{x}$ can be expressed by:

$$F(\boldsymbol{\alpha}) \equiv F_x(\boldsymbol{\alpha}) \equiv \mathscr{P}\{\boldsymbol{x}(\omega) \leqslant \boldsymbol{\alpha}\} = \mathscr{P}\{\omega \in \Omega : \boldsymbol{x}(\omega) \leqslant \boldsymbol{\alpha}\}$$
$$\boldsymbol{\alpha} \in R^n \tag{1.20}$$

in which the distribution function $F(\boldsymbol{\alpha})$ is the probability that the random vector $\boldsymbol{x}$ is less than or equal to a given vector $\boldsymbol{\alpha}$ in an n-dimensional real number field. One can think of the vector $\boldsymbol{x}$ as an

n-dimensional random variable $(x_1, x_2, \ldots, x_n)$ and similarly for $\boldsymbol{\alpha} = (\alpha_1, \alpha_2, \ldots, \alpha_n)$. In this manner the distribution function can also be written as:

$$F_{x_1, x_2, \ldots, x_n} = \mathscr{P}\{x_1 \leqslant \alpha_1, x_2 \leqslant \alpha_2, \ldots, x_n \leqslant \alpha_n\} \tag{1.21}$$

which is the "joint distribution" function of the random variable $(x_1, \ldots, x_n)$. In the two-dimensional case, for example:

$$\begin{aligned} F_{x_1, x_2}(\alpha_1, \alpha_2) &= \mathscr{P}\{x_1 \leqslant \alpha_1, x_2 \leqslant \alpha_2\} \\ F_{x_1, x_2}(\alpha_1, \alpha_2) &= \int_{-\infty}^{\alpha_1} \int_{-\infty}^{\alpha_2} p(\gamma_1, \gamma_2) \mathrm{d}\gamma_1 \mathrm{d}\gamma_2 \end{aligned} \tag{1.22}$$

where $p(\gamma_1, \gamma_2) = p_{x_1, x_2}(\gamma_1, \gamma_2)$. The marginal distribution functions $F(\alpha_1)$, $F(\alpha_2)$ for the continuous case only are obtained from F_{x_1, x_2} as follows:

$$\begin{aligned} F(\alpha_1) &= F_{x_1}(\alpha_1) = F_{x_1, x_2}(\alpha_1, \infty) = \int_{-\infty}^{\alpha_1} p_1(\gamma_1) \mathrm{d}\gamma_1 \\ F(\alpha_2) &= F_{x_2}(\alpha_2) = F_{x_1, x_2}(\infty, \alpha_2) = \int_{-\infty}^{\alpha_2} p_2(\gamma_2) \mathrm{d}\gamma_2 \end{aligned} \tag{1.23}$$

where

$$\begin{aligned} p_1(\gamma_1) &= \int_{-\infty}^{\infty} p(\gamma_1, \gamma_2) \mathrm{d}\gamma_2 \\ p_2(\gamma_2) &= \int_{-\infty}^{\infty} p(\gamma_1, \gamma_2) \mathrm{d}\gamma_1 \end{aligned} \tag{1.24}$$

In general, the "joint density function" can be defined for the continuous case by

$$F(\alpha_1, \ldots, \alpha_n) = \int_{-\infty}^{\alpha_1} \ldots \int_{-\infty}^{\alpha_n} p(\gamma_1, \ldots, \gamma_n) \mathrm{d}\gamma_1, \ldots, \mathrm{d}\gamma_n \tag{1.25}$$

i.e.

$$p(\alpha_1, \ldots, \alpha_n) \equiv \frac{\partial^n F(\alpha_1, \ldots, \alpha_n)}{\partial \alpha_1 \ldots \partial \alpha_n} \qquad \text{for } (\alpha_1, \ldots, \alpha_n) \in R^n$$

The "marginal distribution functions" $F(\alpha_i)$, $i = 1, \ldots, n$ can now be defined as follows:

$$F(\alpha_i) = \int_{-\infty}^{\alpha_i} p_i(\gamma_i)\mathrm{d}\gamma_i$$

$$p_i(\gamma_i) = \underbrace{\int_{-\infty}^{\infty} \dots \int_{-\infty}^{\infty}}_{n-1} p(\gamma_1, \dots, \gamma_n)\mathrm{d}\gamma_1 \dots \mathrm{d}\gamma_{i-1}\mathrm{d}\gamma_{i+1} \dots \mathrm{d}\gamma_n \tag{1.26}$$

(iv) *Conditional distribution and density functions*

The conditional distribution function of two random variables x_1, and x_2 denoted by $F_{x_1|x_2}(\alpha_1|\alpha_2)$ or briefly by $F(\alpha_1|\alpha_2)$ is defined by:

$$F_{x_1|x_2}(\alpha_1|\alpha_2) = F(\alpha_1|\alpha_2) = \mathscr{P}\{x_1 \leqslant \alpha_1|x_2 \leqslant \alpha_2\}$$
$$\forall \alpha_1, \alpha_2 \in R \tag{1.27}$$

where

$$\mathscr{P}\{x_1 \leqslant \alpha_1|x_2 \leqslant \alpha_2\} = \frac{\mathscr{P}\{x_1 \leqslant \alpha_1, x_2 \leqslant \alpha_2\}}{\mathscr{P}\{x_2 \leqslant \alpha_2\}}$$
$$= \frac{F_{x_1,x_2}(\alpha_1, \alpha_2)}{F_{x_2}(\alpha_2)} \tag{1.28}$$

Similarly the conditional distribution function of two random vectors $\boldsymbol{x}_1, \boldsymbol{x}_2$ is defined by:

$$F_{x_1|x_2}(\boldsymbol{\alpha}_1|\boldsymbol{\alpha}_2) = F(\boldsymbol{\alpha}_1|\boldsymbol{\alpha}_2) = \mathscr{P}\{\boldsymbol{x}_1 \leqslant \boldsymbol{\alpha}_1|\boldsymbol{x}_2 \leqslant \boldsymbol{\alpha}_2\} \tag{1.29}$$

which is also

$$\frac{F_{x_1,x_2}(\boldsymbol{\alpha}_1, \boldsymbol{\alpha}_2)}{F_{x_2}(\boldsymbol{\alpha}_2)} \quad \text{or simply} \quad \frac{F(\boldsymbol{\alpha}_1, \boldsymbol{\alpha}_2)}{F(\boldsymbol{\alpha}_2)} \quad \forall \boldsymbol{\alpha}_1, \boldsymbol{\alpha}_2 \in R^n \tag{1.30}$$

Finally, the conditional density function of two random variables is defined as:

$$p_{x_1|x_2}(\alpha_1|\alpha_2) = p(\alpha_1|\alpha_2) = \frac{p(\alpha_1, \alpha_2)}{p(\alpha_2)} \tag{1.31}$$

and that of two random vectors, by:

$$p_{x_1|x_2}(\boldsymbol{\alpha}_1|\boldsymbol{\alpha}_2) = p(\boldsymbol{\alpha}_1|\boldsymbol{\alpha}_2) = \frac{p(\boldsymbol{\alpha}_1, \boldsymbol{\alpha}_2)}{p(\boldsymbol{\alpha}_2)} \tag{1.32}$$

In general, a chain rule of conditional probabilities can be expressed by:

$$p(\alpha_1, \alpha_2, \dots, \alpha_n) = p(\alpha_n)\prod_{i=1}^{n-1} p(\alpha_i|\alpha_{i+1}, \dots, \alpha_n) \tag{1.33}$$

This general relation is useful in the theory of Markov processes, where conditional probability density and distribution functions of several vector components are encountered.

(C) *Functions of random variables*

In the foregoing paragraphs the concepts of probability and of a random variable have been briefly discussed. It has been shown that such a variable can be characterized by its distribution or density function, respectively. A more complete description of a random variable is however achieved by considering first and higher order moments, which are functions of such variables. In another form of representation use is often made of "spectral analysis", the basis of which lies in the concept of the "characteristic function" of a random variable and which is the Fourier transform of its density function. In dealing with discrete interacting systems as discussed later, functions of type $x(\omega)\,\delta(\cdot)$ and $x(\omega)\mathscr{H}(\cdot)$ are employed, where $\delta(\cdot)$ and $\mathscr{H}(\cdot)$ are the Dirac delta and Heaviside step functions, respectively. These types of functions belong to the class of "generalized functions" used later in the text and which will be briefly discussed in the next paragraph.

(i) *Functions of a random variable*

In the application of probabilistic micromechanics one has frequently to deal with the mapping of one set of random variables into another set by means of some functional relationship. Thus consider a function "f" that maps R into R and a random variable x, which by definition maps Ω into R. Assuming that f is at least defined on the range of x one can express a simple relation for any outcome $\omega \in \Omega$ by:

$$g(\omega) = f[x(\omega)] \tag{1.34}$$

such that $f\colon R \to R$ and $g\colon \Omega \to R$.

However, the function $g(\omega)$ in general is not a random variable unless certain restrictions are imposed on the function $f[x(\omega)]$. In particular, the set $\{g \leqslant \beta\}$ must be an event for all $\beta \in R$ and the events $\{g = -\infty\}$ must have zero probability distribution and $\{g = +\infty\}$ a probability distribution one. In order to relate these requirements, let A_β be a set of $x(\omega)$ such that $f[x(\omega)] \leqslant \beta$, i.e.:

$$A_\beta = \{x(\omega); f[x(\omega)] \leqslant \beta\}$$

Hence if the set of outcomes $\{\omega: g(\omega) \leqslant \beta\}$ should be equal to $\{\omega: x(\omega) \in A_\beta\}$, one has:

$$\{\omega: g(\omega) \leqslant \beta\} = \{\omega; x(\omega) \in A_\beta\}$$

or more compactly

$$\{g \leqslant \beta\} = \{x \in A_\beta\}$$

Thus, if $\{g \leqslant \beta\}$ is to be an event, it is necessary that the function "f" be defined so that $\{x \in A_\beta\}$ is an event for $\forall \beta \in R$. This restriction together with the condition $\mathscr{P}\{g(-\infty)\} = 0$; $\mathscr{P}\{g(\infty)\} = 1$ makes then $g(\omega)$ a random variable. In order to find the distribution and density function of g, one can use the basic definitions of these functions related to the probability space, i.e. $g \in (\Omega, \mathscr{F}, \mathscr{P})$ and $g(\omega) \in R$. If the distribution of the random variable x is known, then by using $f(x) \rightarrow g$ one can also establish the distribution of g. There are two cases to distinguish, i.e. the

(a) *Discrete case*

If x is a discrete random variable, then the density function is given by:

$$p_x(\alpha) = \sum_{i=1}^{n} \varrho_i \, \delta(\alpha - \alpha_i), \qquad n \text{ finite or infinite} \tag{1.35}$$

in which

$$\varrho_i = \mathscr{P}\{x = \alpha_i\} \tag{1.36}$$

Assuming now that the function f is one-to-one, i.e. that for every $\alpha_i \in R$ there exists a corresponding $\beta_i \in R$ such that $f(\alpha_i) = \beta_i$ one can express the density function as follows:

$$p_g(\beta) = \sum_{i=1}^{n} \Phi_i \, \delta(\beta - \beta_i)$$

where Φ_i is the probability $\mathscr{P}\{g = \beta_i\}$. However, on the assumption that f is one-to-one and $\{g \leqslant \beta\} = \{x \in A_\beta\} = \{x \leqslant \alpha\}$ it follows that:

$$\mathscr{P}\{g = \beta_i\} = \mathscr{P}\{x = \alpha_i\} \Rightarrow \Phi_i = \varrho_i$$

and thus

$$p_g(\beta) = \sum_{i=1}^{n} \varrho_i \delta(\beta - \beta_i) \tag{1.37}$$

being the density function of $g(\omega)$.

(b) *Continuous case*

As shown earlier the probability density in the case of a continuous random variable is defined by:

$$F_x(\alpha) = \int_{-\infty}^{\alpha} p_x(\gamma)\mathrm{d}\gamma = \mathscr{P}\{x \leqslant \alpha\}$$

Analogously one can define the density function of $g(\omega)$ or $g(x)$ in terms of its distribution function as follows:

$$F_g(\beta) = \int_{-\infty}^{\beta} p_g(\lambda)\mathrm{d}\lambda = \mathscr{P}\{g \leqslant \beta\} \tag{1.38}$$

but

$$\mathscr{P}\{g \leqslant \beta\} = \mathscr{P}\{x \in A_\beta\} = \int_{A_\beta} p_x(\gamma)\mathrm{d}\gamma$$

hence

$$F_g(\beta) = \int_{-\infty}^{\beta} p_g(\lambda)\,\mathrm{d}\lambda = \int_{A_\beta} p_x(\gamma)\mathrm{d}\gamma$$

On the assumption that $F_g(\beta)$ is continuously differentiable, then:

$$p_g(\beta) = \frac{\partial F_g(\beta)}{\partial \beta} \tag{1.39}$$

It should be noted that by restricting the class of mapping f one can extract a relation between $p_x(\gamma)$ and $p_g(\lambda)$. This will not be pursued however in this text.

(ii) *Generalized functions*

In view of the importance of discrete random variables in probabilistic micromechanics of structured media and the assessment of their corresponding distribution functions from experimental observations, the concept of "generalized functions" becomes often significant. This concept will be briefly discussed in this section. A more comprehensive study is given amongst others in the texts of Schwartz[13], Sneddon[14], Gel'fand and Shilov[15] and Gel'fand and Vilenkin[16]. The considerations given here follow essentially the latter references.

Considering a set of real functions $\{\Phi\} \subset K$, which have continuous derivatives of all orders and which have compact supports (i.e. $\Phi(x)$ vanishes everywhere outside a compact set, e.g. $a \leqslant x \leqslant b$). It is also assumed that the derivatives of Φ belong to K. Functions which have these properties are referred to as "test functions". Test functions $\Phi(x)$ can be added or multiplied by real numbers to yield other test functions. Hence the space K is a linear space. Considering now another function called a "linear functional" if for every test function $\Phi(x) \in K$ there exists a real number in such a manner, that the following conditions hold:

I. for any $\alpha_1, \alpha_2 \in R$ and $\Phi_1, \Phi_2 \in K$
$(\Psi, \alpha_1\Phi_1+\alpha_2\Phi_2) = \alpha_1(\Psi, \Phi_1)+\alpha_2(\Psi, \Phi_2)$ is a real number

II. if the sequence $\Phi_1, \Phi_2, \ldots, \Phi_n$ converges to zero in K, then $(\Psi, \Phi_1), (\Psi, \Phi_2), \ldots, (\Psi, \Phi_n)$ also converges to zero in R

If, for example, the functional $\Psi(x)$ is an integrable function in the Lebesgue sense and is bounded in the n-dimensional space R^n, one can associate every test function $\Phi(x)$ with this functional as follows:

$$(\Psi, \Phi) = \int_{R^n} \Psi(x)\Phi(x)\mathrm{d}x \tag{1.40}$$

where the integral is taken over the bounded region in which $\Phi(x)$ fails to vanish. This form (Ψ, Φ) represents then a special subclass of the "generalized functions". Other types of linear functionals on K exist as well. It is well known that not all "ordinary functions" are differentiable. In fact there exists many functions which have no derivatives in the usual sense. In contrast "generalized functions" are always differentiable and have derivatives of all orders, which again are generalized functions. Considering $\Psi(x)$ to be a piecewise continuous function with piecewise continuous first derivative and expressing the first derivative of the functional in the usual manner, then:

$$(\Psi', \Phi) = \int_{-\infty}^{+\infty} \Psi'(x)\Phi(x)\mathrm{d}x \tag{1.41}$$

which upon integration by parts and since $\Phi(x)$ vanishes outside an interval say $[a, b]$ on K leads to:

$$(\Psi', \Phi) = (\Psi, -\Phi') \tag{1.42}$$

which can be defined as the first derivative of a generalized function or a continuous linear functional Ψ by the form:

$$(\Psi, -\Phi') = \Psi' \quad \text{or} \quad \frac{\mathrm{d}\Psi}{\mathrm{d}x} \tag{1.43}$$

Using the definition of the linear functional given above the derivative can be expressed by[15,16]:

$$\int_{-\infty}^{+\infty} \Psi'(x)\Phi(x)\mathrm{d}x = -\int_{-\infty}^{+\infty} \Psi(x)\Phi'(x)\mathrm{d}x \tag{1.44}$$

The consistency of this relation has been proved by Gel'fand[17]. The above formulation can be extended to the case of several variables. In an analogous form "partial derivatives" of a generalized function can also be defined and since differentiation of a generalized function yields again a generalized function, the process may be continued to obtain partial derivatives of any order.

It has been mentioned earlier that of particular interest in the subsequent analysis are "singular functions" such as for example the Dirac-delta function $\delta(x)$. Strictly speaking $\delta(x)$ is not a function in the classical sense. However, to every such singular function corresponds a linear functional that associates with every well-behaved function some defined number such that:

$$(\delta(x), \Phi(x)) = \Phi(0) \tag{1.45}$$

Sometimes one has also to consider a "translated delta function", i.e. $\delta(x-x_0)$ in which case the functional is defined by:

$$(\delta(x-x_0), \Phi(x)) = \int_{-\infty}^{+\infty} \Phi(x)\mathrm{d}\mu_0(x) = \Phi(x_0) \tag{1.46}$$

where $\mathrm{d}\mu_0(x)$ is a measure concentrated at point x_0 and written in the form $\mathrm{d}\mu_0(x) \equiv \delta(x-x_0)\mathrm{d}x$.

The above concept of generalized functions can now be applied to define the probability density function of a discrete random variable often encountered in experimental micromechanics. For this purpose consider a piecewise continuous distribution function $F_x(\alpha)$ which is obtained from certain experiments and which has piecewise continuous derivatives $p_x(\alpha)$ (Fig. 8, below).

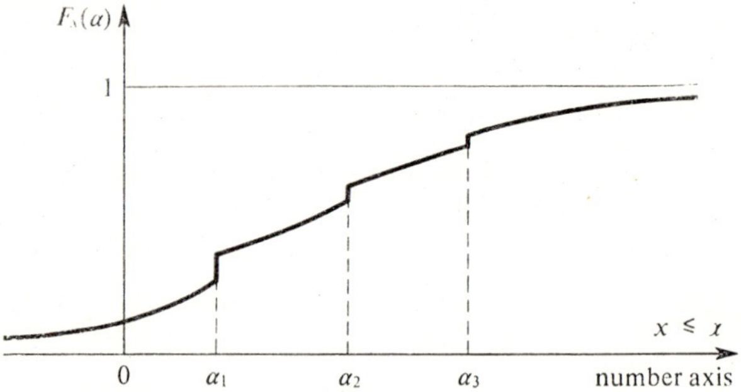

Fig. 8. Piecewise continuous distribution function.

This means that $p_x(\alpha)$ is defined everywhere except at a finite number of discontinuities. Considering now a function of the following form:

$$\mathscr{H} = \begin{cases} 0 & x < 0 \\ 1 & x > 0 \end{cases} \tag{1.47}$$

it follows that

$$\begin{aligned} (\mathscr{H}'(x), \Phi(x)) &= (\mathscr{H}(x), -\Phi'(x)) \\ &= -\int_0^\infty \Phi'(x)\mathrm{d}x = \Phi(0) \end{aligned} \tag{1.48}$$

In accordance with the earlier definition (1.43) that the derivative of the linear functional $\mathscr{H}(x)$, i.e. $\mathscr{H}'(x)$ is given by:

$$\mathscr{H}'(x) = \delta(x)$$

and also

$$\mathscr{H}'(x-x_k) = \delta(x-x_k)$$

Identifying x_k by α_k on the α-number axis, then the last relation becomes:

$$\mathscr{H}'(\alpha-\alpha_k) = \delta(\alpha-\alpha_k) \tag{1.49}$$

where α_k is the position of the discontinuity. If "j_k" denotes the height of the jump at the kth discontinuity, then by defining a function of the form:

$$\mathscr{F}_x(\alpha) = F_x(\alpha) - \sum_{k=1}^{n} j_k \mathscr{H}(\alpha - \alpha_k) \tag{1.50}$$

it is seen that this function is continuous everywhere and is a regular generalized function. Thus it is differentiable and has a continuous derivative $f_x(\alpha)$ such that:

$$f_x(\alpha) = p_x(\alpha) - \sum_{k=1}^{n} j_k \delta(\alpha - \alpha_k) \tag{1.51}$$

from which it follows that:

$$p_x(\alpha) = f_x(\alpha) + \sum_{k=1}^{n} j_k \delta(\alpha - \alpha_k) \tag{1.52}$$

In particular, if the jump height j_k is replaced by the discrete probability ϱ_k, then:

$$p_x(\alpha) = f_x(\alpha) + \sum_{k=1}^{n} \varrho_k \delta(\alpha - \alpha_k) \tag{1.53}$$

Hence it is seen that, if the distribution function of a discrete random variable is a piecewise continuous function having a finite number of discontinuities, the corresponding probability density function can be expressed in terms of a continuous differentiable functional $f(x)$ and a linear combination of δ-functions as shown above. The above argument may be generated to include the case of a series of discrete random variables and for spaces of any dimensions.

(iii) *Other functions of random variables*

In dealing with field quantities that are random variables or functions of such variables, one of the main functions of a random variable is its "expected value" or first moment.

(a) *Expected values*

The expected value of a continuous random variable x is denoted by $E\{x\}$ and is given by:

$$E\{x\} = \int_{-\infty}^{+\infty} \alpha p_x(\alpha)\,\mathrm{d}\alpha, \qquad x(\omega) \leqslant \alpha \tag{1.54}$$

if this integral exists. In the discrete case as shown above, the density function can be simplified (eqn. 1.35) to:

$$p_x(\alpha) = \sum_{i=1}^{n} \varrho_i \delta(\alpha - \alpha_i), \qquad \varrho_i = \mathscr{P}\{x = \alpha_i\}$$

so that the expected value of the discrete random variable becomes:

$$E\{x\} = \sum_{i=1}^{n} \varrho_i \alpha_i = \sum_{i=1}^{n} \alpha_i \mathscr{P}\{x = \alpha_i\} \tag{1.55}$$

(b) *Statistical moments "m"*

Whilst the expected value or first integral moment of a random variable gives its average or "mean value", higher order statistical moments are significant since they describe the shape of the density curve in particular at the ends of the number axis. This is illustrated in Fig. 9 below. From probability theory, the integral moment of order n for a continuous r.v. is defined as follows:

$$m^{(n)} = E\{x^n\} = \int_{-\infty}^{+\infty} \alpha^n p_x(\alpha) \mathrm{d}\alpha \tag{1.56}$$

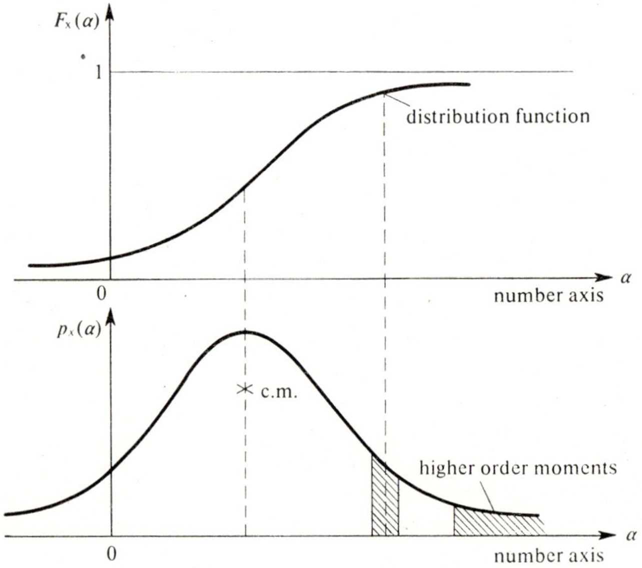

Fig. 9. Distribution and density functions of a continuous r.v.

Hence the first order moment is seen to be:

$$m^{(1)} = E\{x\} = \int_{-\infty}^{+\infty} \alpha p_x(\alpha)\mathrm{d}\alpha$$

which is the same as given in eqn. (1.54).

The definition of the nth order integral moment holds also for any arbitrary function of a random variable. Thus:

$$m_g^{(n)} = E\{g(x)^n\} = \int_{-\infty}^{+\infty} \beta^n p_g(\beta)\mathrm{d}\beta$$

$$n = 1, 2, \ldots \qquad g(x) \leqslant \beta \tag{1.57}$$

so that the expected value of this function is expressed by:

$$m_g^{(1)} = E\{g(x)\} = \int_{-\infty}^{+\infty} \beta p_g(\beta)\mathrm{d}\beta$$

It is to be noted that the above definition of statistical moments can be extended to generalized functions $\Psi(x)$ associated with a test function $\Phi(x) \in K$. However, in this case one has to deal with a generalized distribution function. In this manner:

$$m_\Psi^{(n)} = E\{\Psi(x)^n\} = \int_{-\infty}^{+\infty} \Psi^n(x)\mathrm{d}\mathscr{F}(x)$$

$$= \int_{-\infty}^{+\infty} \Psi^n(x)f(x)\mathrm{d}x \tag{1.58}$$

and in which the expected value is of the form:

$$m_\Psi^{(1)} = E\{\Psi(x)\} = \int_{-\infty}^{+\infty} \Psi(x)f(x)\mathrm{d}x \tag{1.59}$$

Considering now that the variable $x(\omega)$ is a generalized function, then

$$m^{(n)} = E\{x^n\} = \int_{-\infty}^{+\infty} \alpha^n \mathrm{d}\mathscr{F}(x) = \int_{-\infty}^{+\infty} \alpha^n f(\alpha)\mathrm{d}\alpha \tag{1.60}$$

where $f(\alpha)$ is the continuous linear functional and is called the *generalized density function* of x.

(c) *Central moments "M"*

In probability theory the "central moments" of a random quantity are sometimes significant. Thus the definition of these moments is as follows:

$$M^{(n)} = E\{(x-m^{(1)})^n\}$$
$$= \int_{-\infty}^{+\infty} (\alpha-m^{(1)})^n p_x(\alpha)\,d\alpha, \quad n = 1, 2, \ldots \tag{1.61}$$

Alternatively the definition of the nth order central moment can be given as:

$$M^{(n)} = \int_{-\infty}^{+\infty} (\alpha-m^{(1)})^n dF_x(\alpha)$$
$$= \int_{-\infty}^{+\infty} (\alpha-E\{x\})^n p_x(\alpha)\,d\alpha \tag{1.62}$$

For the case of a scalar random variable, it is seen that $M^{(1)} = 0$, since

$$M^{(1)} = \int_{-\infty}^{+\infty} (\alpha-E\{x\}) p_x(\alpha)\,d\alpha$$
$$= \int_{-\infty}^{+\infty} (\alpha-m^{(1)})\, p_x(\alpha)\,d\alpha = m^{(1)}-m^{(1)} = 0 \tag{1.63}$$

In many practical applications involving distribution and density functions of a random variable, the second central moment is even more significant. It is also referred to as the "variance" and is given by the following expression:

$$M^{(2)} = E\{(x-m^{(1)})^2\} = E\{[x-E\{x\}]^2\}$$
$$= \int_{-\infty}^{+\infty} (\alpha-m^{(1)})^2 p_x(\alpha)\,d\alpha \tag{1.64}$$

The square root of the variance is known as the "standard deviation" and is usually denoted by the symbol σ. It should be noted that these simple definitions can be readily extended to higher dimensions for the case of random vectors or random tensors.

Thus, if x and y are two random variables, which may be regarded as components of a random vector $\boldsymbol{x}$, "mixed moments" will arise, which are defined in this case of the order $k+l$, as follows:

$$m^{(k+l)} = m_{kl} = E\{x^k y^l\} = \int_{-\infty}^{+\infty}\int_{-\infty}^{+\infty} \alpha^k \beta^l p_{x,y}(\alpha, \beta)\,\mathrm{d}\alpha\,\mathrm{d}\beta \qquad (1.65)$$

in which $x(\omega) \leqslant \alpha$, $y(\omega) \leqslant \beta$ and $p_{x,y}(\alpha, \beta)$ represents the joint probability density function of the two variables. Analogously, the "mixed central moment" of order $k+l$ will be given by:

$$\begin{aligned} M^{(k+l)} &= M_{kl} = E\{[x-m^{(1)}]^k [y-m^{(1)}]^l\} \\ &= \int_{-\infty}^{+\infty}\int_{-\infty}^{+\infty} (\alpha-m)^k (\beta-m)^l p_{x,y}(\alpha, \beta)\,\mathrm{d}\alpha\,\mathrm{d}\beta \end{aligned} \qquad (1.66)$$

The mixed central moment M_{11} is frequently referred to as the "covariance" and is denoted by the symbol k_{xy}[7]. Often a "correlation coefficient" is used instead of the covariance function. This coefficient in the case of a two-dimensional random vector with components x, y is defined as:

$$\varrho_{xy} = \frac{k_{xy}}{\sqrt{M_x^{(2)} M_y^{(2)}}} = \frac{k_{xy}}{\sigma_x \sigma_y} \qquad (1.67)$$

It should be noted that in the case of a two-dimensional vector written as x_i, $(i = 1, 2)$ the covariance k_{xy} forms a matrix k_{ij}, $(i = 1, 2)$, which by extending the argument above to the n-dimensional case yields a $n \times n$ matrix k_{ij} to represent the covariance as follows:

$$k = [k_{ij}] = \begin{bmatrix} k_{11} & \ldots & k_{1n} \\ \vdots & & \vdots \\ k_{n1} & \ldots & k_{nn} \end{bmatrix} \qquad (1.68)$$

Similarly one can write a "correlation matrix" for the n-dimensional system in the form of:

$$\varrho = [\varrho_{ij}] = \begin{bmatrix} \varrho_{11} & \ldots & \varrho_{1n} \\ \vdots & & \vdots \\ \varrho_{n1} & \ldots & \varrho_{nn} \end{bmatrix} \qquad (1.69)$$

Finally, this correlation matrix can be normalized in the following manner:

$$r_{ij} = \frac{\varrho_{ij}}{\sqrt{\varrho_{ii}\varrho_{jj}}} \tag{1.70}$$

so that the "normalized correlation matrix" becomes:

$$R = [r_{ij}] = \begin{bmatrix} 1 & \cdots & r_{1n} \\ \vdots & & \vdots \\ r_{n1} & \cdots & 1 \end{bmatrix} \tag{1.71}$$

1.5 Stochastic processes

(A) *Stochastic functions*

In the foregoing sections random variables and certain relations of the theory of probability were briefly discussed. These quantities assume certain values in any given trial or experiment. In probabilistic micromechanics of solids one often encounters, however, random variables that depend also on other parameters such as time for example. Random variables of this type can assume in the outcome of an experiment an infinite number of values or a set of values which will depend on time as well as on other parameters. One deals therefore in this case with a function that has only one uniquely determined value that corresponds to each instant of time over the interval in which the observations are made.

In other words, if another observation is made under identical physical conditions, a different function will be obtained. Such a function is therefore called a *stochastic* or *random function* and will be denoted by $x_t(\omega)$, $t \in T$ or $x(t, \omega)$. Hence a stochastic function is defined as one for which a particular value of the argument t, $x_t(\omega)$ is a random quantity, i.e. $x(\omega)$. Any function which may become equal to the random function as a result of an observation is referred to as a "realization or sample" function, $x_t(\omega)$.

A random function may be regarded as specified, if for each element of its argument and $t \in T$, its distribution function can be expressed by:

$$F_t(\alpha) = F_x(\alpha, t) = \mathscr{P}\{x(t) \leqslant \alpha\}, \quad t \in T \tag{1.72}$$

Similar to the case of a two-dimensional random vector as shown earlier, one can write for $x(t_1, t_2) = [x(t_1), x(t_2)]$ the joint distribution function as follows:

$$\begin{aligned} F_{t_1,t_2}(\alpha_1, \alpha_2) &= F_x(\alpha_1, t_1; \alpha_2, t_2) \\ &= \mathscr{P}\{x(t_1) \leqslant \alpha_1, x(t_2) \leqslant \alpha_2\} \end{aligned} \tag{1.73}$$

Extending this argument to the n-dimensional case, then in general:

$$x(t_1, t_2, \ldots, t_n) = [x(t_1), x(t_2), \ldots, x(t_n)] \tag{1.74}$$

and the corresponding distribution function is then:

$$F_{t_1,t_2,\ldots,t_n}(\alpha_1, \alpha_2, \ldots, \alpha_n) = \mathscr{P}\{x(t_1) \leqslant \alpha_1, \ldots, x(t_n) \leqslant \alpha_n\} \tag{1.75}$$

It should be noted that since $x(t) = x_t$ is a random function of time, there will be for every finite set $t_1, t_2, \ldots, t_n$ a set of random variables $x_1, x_2, \ldots, x_n$ so that $x_i = x(t_i)$ and hence the distribution function can be written in general as follows:

$$\begin{aligned} F_{t_i}(\alpha_i) &\equiv F_{t_1,\ldots,t_i}(\alpha_1, \ldots, \alpha_i) \\ &= F_{t_1,\ldots,t_n}(\alpha_1, \ldots, \alpha_i, \infty, \ldots, \infty) \\ &= \int_{-\infty}^{\alpha_1} \cdots \int_{-\infty}^{\alpha_i} p_{t_1,\ldots,t_i}(\gamma_1, \ldots, \gamma_i)\mathrm{d}\gamma_1, \ldots, \mathrm{d}\gamma_i, \end{aligned} \tag{1.76}$$

$$p_{t_1,\ldots,t_i}(\alpha_1,\ldots,\alpha_i) = \underbrace{\int_{-\infty}^{\infty} \cdots \int_{-\infty}^{\infty}}_{n-i} p_{t_1,\ldots,t_n}(\alpha_1,\ldots,\alpha_n)\mathrm{d}\gamma_{i+1},\ldots,\mathrm{d}\gamma_n$$

from which formally the probability density can be expressed by

$$\begin{aligned} p_{t_i}(\alpha_i) &\equiv p_{t_1,\ldots,t_i}(\alpha_1, \ldots, \alpha_i) \\ &= \frac{\partial^i F_{t_1,\ldots,t_i}(\alpha_1, \ldots, \alpha_i)}{\partial\alpha_1 \ldots \partial\alpha_i} \equiv \frac{\partial^i F_{t_i}(\alpha_i)}{\partial\alpha_i^i} \end{aligned} \tag{1.77}$$

The distribution function as stated in equation (1.73) can be shown to satisfy the symmetry condition and the compatibility condition for any set $t_{i+1} \ldots t_n$ provided $i < n$. It is also apparent that the probability density is non-negative, i.e. $p_{t_i}(\alpha_i) \geqslant 0$ and that its integration leads to one.

The set $t \in T$ often consists of a whole set of real numbers going from $-\infty$ to $+\infty$. In this case $x(t)$ is regarded as a "random process".

If, however, T is a set of integers, then one has a "random sequence". The latter is frequently encountered in dealing with the analysis of temporal discrete systems. A distinct property of a random function may be its "stationarity", e.g. that all finite dimensional distribution functions that define $x(t)$ remain the same, if the whole set $t \in T$, $t_1, \dots, t_n$ is shifted along the time axis. Thus for all "stationary random functions" the one-dimensional distribution is the same, whilst the two-dimensional distribution depends only on the difference $t_2 - t_1$, etc.

It should be noted that the concept of a stochastic function briefly discussed above can be readily extended to vectors and tensors. A random tensor for example may be regarded as one, whose elements are stochastic variables or stochastic functions. In general, any mathematical quantity in the deterministic formulation has an analogous form in the stochastic description. For a more comprehensive study the reader is referred to several texts as listed in the bibliography, in particular to the work of Pugachev[7].

(B) *Some properties of random functions*

In the analysis of stochastic functions various theorems are required, in particular those relating to convergence, continuity, etc. Such properties of random functions become important when dealing with the differentiation or integration of a random function. Furthermore, the knowledge of convergence and continuity are necessary for the definition of "stationarity" of a random process.

For this purpose considering a random sequence $x_1(\omega)$, $x_2(\omega)$, ..., $x_n(\omega)$ of a random variable $x(\omega)$, $\omega \in \Omega$, then the following definitions for convergence can be given:

(i) *Convergence*

(a) *e-convergence* (convergence everywhere)

$$x_n(\omega) \to x(\omega)$$

if

$$\lim_{n\to\infty} x_n(\omega) = x(\omega) \quad \text{for} \quad \forall\omega \in \Omega \tag{1.78}$$

(b) *a.e.-convergence* (convergence almost everywhere)

$$x_n(\omega) \to x(\omega)$$

if

$$\mathscr{P}\{\lim_{n\to\infty} x_n(\omega) = x(\omega)\} = 1 \tag{1.79}$$

The "a.e.-convergence" is sometimes called "convergence with probability 1".

(c) *p-convergence* (convergence in the probability)

$$x_n(\omega) \to x(\omega)$$

if

$$\lim_{n\to\infty} \mathscr{P}\{|x_n - x| \geqslant \varepsilon\} = 0 \quad \text{for} \quad \forall\varepsilon > 0 \tag{1.80}$$

(d) *mean-convergence*

$$x_n(\omega) \xrightarrow{\text{mean}} x(\omega)$$

if

$$\lim_{n\to\infty} E\{|x_n - x|\} = 0 \tag{1.81}$$

One can also consider a "mean-convergence" in the form

$$\lim_{n\to\infty} \frac{1}{n} \sum_{i=1}^{n} |x_i - x^*| = 0 \tag{1.82}$$

where $x^* = x^*(\omega)$, $\omega \in \Omega$ is a random variable. The convergence in (1.82) can be understood in many ways; the formulae (1.81) and (1.82) are not equivalent: in general, $x^* \neq x$. If $\forall\omega \in \Omega$, $x^*(\omega) = Ex$, then we say that the sequence of random variables x_n has the "ergodicity property" (see B, (iv), p. 40). It is seen from the definition of mean-convergence that every Cauchy sequence $\{x_n(\omega)\}$ converges in the mean since for x_n and x_m

$$E\{|x_n - x_m|\} \leqslant E\{|x_n - x|\} + E\{|x_m - x|\} \tag{1.83}$$

and hence

$$\lim_{\substack{n\to\infty \\ m\to\infty}} E\{|x_n - x_m|\}$$

$$\leqslant \lim_{n\to\infty} E\{|x_n - x|\} + \lim_{m\to\infty} E\{|x_m - x|\} = 0 \tag{1.84}$$

(e) *m.-s.-convergence* (convergence in the mean-square)

$$x_n(\omega) \xrightarrow{\text{m-s}} x(\omega) \tag{1.85}$$

if

$$\lim_{n\to\infty} E\{|x_n - x|^2\} = 0 \tag{1.86}$$

The counterpart of (1.82) is here "convergence in the mean-square" of the form

$$\lim_{n\to\infty} \frac{1}{n}\sum_{i=1}^{n} |x_i - x^*|^2 = 0 \tag{1.87}$$

The remarks following the formula (1.82) are also valid for (1.87). In particular, for the case $x^* \neq x$.

The definitions (a)–(e) of the convergence are not equivalent, e.g. it is possible for a random sequence to converge almost everywhere, but not in the mean-square. The above definitions of convergence given briefly here are equally valid for stochastic functions. Thus considering a sequence of random functions $x(t_1) = x_1, x(t_2) = x_2, \ldots$ and so on, then one says that $x(t_n) = x_n$ converges to x in the probability, when t_n converges to t, if

$$\lim_{n\to\infty} \mathscr{P}\{|x(t_n) - x(t)| \geqslant \varepsilon\} = 0 \quad \text{for} \quad \forall\varepsilon > 0 \tag{1.88}$$

and if $t_1, t_2, \ldots, t_n \to t$ for $n \to \infty$.

Similarly in the mean-square sense $x(t_n) \to x(t)$, if

$$\lim_{n\to\infty} x(t_n) = x(t) \quad \text{whenever} \quad t_n \to t \text{ for } n \to \infty \tag{1.89}$$

(ii) *Continuity*

As shown in the foregoing section several convergence criteria for a random function or a random process are possible, and hence there exist accordingly several definitions of continuity. However, only those important in the later analysis will be given briefly below.

(a) *p-continuity* (continuity in probability)

Consider the random function x with realizations $x(t)$ and $x(t+s)$, where $s, t \in T$. Then $x(t)$ is said to be continuous at t in the probability, if

$$\lim_{s\to 0} \mathscr{P}\{|x(t+s) - x(t)| \leqslant \varepsilon\} = 1 \quad \text{for} \quad \forall\varepsilon > 0 \tag{1.90}$$

or equivalently, if

$$\lim_{s\to 0} \mathscr{P}\{|x(t+s) - x(t)| > \varepsilon\} = 0 \quad \text{for} \quad \forall\varepsilon > 0 \tag{1.91}$$

One can say that every Cauchy sequence $\{x(t_n); n = 1, 2, \ldots\}$ converges in the probability.

(b) *m-s-continuity* (continuity in the mean-square)

Again the random function $x(t)$ is said to be continuous in the mean-square sense at t, if

$$\lim_{s \to 0} E\{|x(t+s)-x(t)|^2\} = 0 \tag{1.92}$$

Analogously to the p-continuity, the mean-square continuity implies that every Cauchy sequence $\{x(t_n);\ n = 1, 2, \ldots\}$ converges in the mean-square.

(iii) *Differentiation and integration of stochastic functions*

It has been stated earlier that, in general, the theorems applicable to ordinary differential and integral calculus remain valid also in the case of random quantities, but require certain modifications. These modifications originate from the theory of probability and the necessity of defining an appropriate probability measure. There also exists as indicated in the foregoing section a correspondance between the convergence principles concerning random variables and those valid for stochastic functions. One concept however is new, e.g. the interchangeability between the operation of taking the expected value of a random quantity and that of its differentiation. Thus a stochastic function $x(t)$ is said to have a *mean-square derivative* $x'(t)$ written as:

$$x'(t) = \frac{\mathrm{d}}{\mathrm{d}t}(x(t)) \tag{1.93}$$

where $\frac{\mathrm{d}}{\mathrm{d}t}$ represents a mean square differential operator, if $x(t)$ converges in the mean-square to $x'(t)$, i.e.:

$$\lim_{\Delta t \to 0} E\left\{\left[\frac{x(t+\Delta t)-x(t)}{\Delta t} - x'(t)\right]^2\right\} = 0 \tag{1.94}$$

A necessary and sufficient condition that $x(t)$ has a mean-square derivative at t is that the derivative:

$$\frac{\partial^2}{\partial t_1 \partial t_2} E\{x(t_1)x(t_2)\} \quad \text{exists at} \quad (t_1, t_2) = (t, t)$$

If $x(t)$ has a mean-square derivative for every $t \in T$, then

$$\frac{\partial}{\partial t_1} E\{x(t_1)x(t_2)\}, \quad \frac{\partial}{\partial t_2} E\{x(t_1)x(t_2)\} \quad \text{for} \quad \forall t_1, t_2 \in T$$

exist as well.

Consider the interval $[a, b] \subset T$ and divide it into n parts by means of the points $a = t_0 < t_1 < t_2 < \ldots < t_n = b$. Let $\xi_i \in [t_i, t_{i+1}]$ and construct the approximate integral sum

$$I_n(x) = \sum_{i=0}^{n-1} x(\xi_i)\,(t_{i+1} - t_i) \tag{1.95}$$

We say that the stochastic function $x(t)$ is *integrable* in the considered sense of convergence (see the definitions in B, (iii)), if the approximate sums in (1.95) have a limit $I(x)$:

$$\lim_{n\to\infty} I_n(x) = I(x) \tag{1.96}$$

in the considered sense of convergence, independently of the division of the interval $[a, b]$ and the choice of the points ξ_i. The limit $I(x)$ entering (1.96) is written in the form

$$I(x) = \int_a^b x(t)\,\mathrm{d}t \tag{1.97}$$

If the convergence in (1.96) is the convergence everywhere, then we simply say that the stochastic function $x(t)$ is integrable. In a generalized sense such an integral will also exist, such that:

$$\begin{aligned} &\lim_{n\to\infty} I_n(\Phi x) = \int_a^b \Phi(t)x(t)\,\mathrm{d}t \\ &(\Phi x)(t) = \Phi(t)x(t), \quad \forall t \in [a, b] \end{aligned} \tag{1.98}$$

in which the test function $\Phi(t)$ as discussed previously is considered to be a continuous scalar function with a compact support $[a, b]$. Before dealing with the rather important property of a stochastic function, i.e. its "stationarity", for completeness of this review the distribution functions and statistical moments of a stochastic process should be mentioned. Thus, the n-dimensional distribution function of a stochastic process $x(t)$ can be expressed by:

$$\begin{aligned} &F_{t_1, t_2, \ldots, t_n}(\alpha_1, \alpha_2, \ldots, \alpha_n) \\ &= \mathscr{P}\{x(t_1) \leqslant \alpha_1, \ldots, x(t_n) \leqslant \alpha_n\} \end{aligned} \tag{1.99}$$

and hence the corresponding density function for a continuous process becomes:

$$p_{t_1, t_2, \dots, t_n}(\alpha_1, \alpha_2, \dots, \alpha_n) = \frac{\partial^n F_{t_1, t_2 \dots, t_n}(\alpha_1, \alpha_2, \dots, \alpha_n)}{\partial \alpha_1 \partial \alpha_2 \dots \partial \alpha_n} \tag{1.100}$$

The n-dimensional moment can be stated as follows:

$$\mu_{r_1 \dots r_n}(t_1, \dots, t_n) = E\{[x(t_1)]^{r_1} \dots [x(t_n)]^{r_n}\} = \underbrace{\int_{-\infty}^{+\infty} \dots \int_{-\infty}^{+\infty}}_{n} \alpha_1^{r_1} \dots \alpha_n^{r_n} \mathrm{d}F_{t_1, \dots, t_n}(\alpha_1, \dots, \alpha_n) \tag{1.101}$$

representing an n-fold Stieltjes integral. It follows that the first moment is given by:

$$\mu_1(t) = E\{x(t)\} = \int_{-\infty}^{+\infty} \alpha \mathrm{d}F_t(\alpha) \tag{1.102}$$

and the second moment by:

$$\mu_{11}(t_1, t_2) = R(t_1, t_2) = E\{[x(t_1)]\,[x(t_2)]\} = \int_{-\infty}^{+\infty}\int_{-\infty}^{+\infty} \alpha_1 \alpha_2 \mathrm{d}F_{t_1, t_2}(\alpha_1, \alpha_2) \tag{1.103}$$

(iv) *Stationarity*

By considering a random process in terms of the index set $t \in T$, one can distinguish between a continuous or discrete process. However, a more important distinction of such a process is whether it is a "strictly stationary" or a "weakly stationary" one. The corresponding definitions are given below.

(a) *Strict stationarity*

A stochastic process $x(t)$ will be called "strictly stationary", if the n-dimensional distribution function remains the same for a time shift along the T-axis, i.e. if:

$$F_{t_1, \dots, t_n}(\alpha_1 \dots \alpha_n) = F_{t_1+\tau, \dots, t_n+\tau}(\alpha_1, \dots, \alpha_n) \quad \text{for} \quad \forall \tau \in T \tag{1.104}$$

This implies that $F_t(\alpha)$ is independent of t and $F_{t_1, t_2}(\alpha_1, \alpha_2)$ depends only on the difference $t_2 - t_1$, for $t_2 > t_1$.

(b) *Weak or wide stationarity*

A stochastic process $x(t)$ is called "weakly stationary" or "stationary in the wide sense", if there exist the first and second moments of $x(t)$ (see (1.102), (1.103)) and

$$\begin{aligned} \mu_1(t) &= E\{x(t)\} = \mu_1 = \text{const} \quad \text{for} \quad \forall t \in T \\ \mu_{11}(t_1, t_2) &= R(t_1, t_2) = R(t_2 - t_1) \quad \text{for} \quad \forall t_1, t_2 \in T \end{aligned} \tag{1.105}$$

If a stochastic process $x(t)$ is strictly stationary and

$$E\{|x(t)|^2\} = \int_{-\infty}^{\infty} |\alpha|^2 \mathrm{d}F_t(\alpha) < \infty \quad \text{for} \quad \forall t \in T \tag{1.106}$$

then the process $x(t)$ is stationary in the wide sense.

It is of interest to note that these statistics of a random process can be obtained from a "single realization". This approach is usually taken in correlation theory (see Yaglom[8]) and is based on the assumption of the process to be an ergodic one. In Chapter III dealing with the application of functional analysis in micromechanics a brief discussion on the ergodic theorems will be given in a somewhat more general manner than stated below. However, for the present considerations, it may be stated that the "ergodicity" of a process implies that the spatial average of a field quantity, if considered as a stationary stochastic process $x(t)$ and which is associated with the entire space of experimental outcomes Ω at time t, can be replaced by its corresponding time average.

As a consequence the statistical characteristics of a stationary random process $x(t)$ can be expressed as follows:

$$\mu(t_1) = E\{x(t)\} = \mu_1 = \text{const} \quad \text{for} \quad \forall t \in T \tag{1.107}$$

in which for the discrete case

$$\mu_1 = \lim_{n\to\infty} \frac{1}{n+1} \sum_{i=0}^{n} \alpha_i \tag{1.108}$$

for a sequence of random variables of the form $x_i = x(t_i)$, $i = 0, 1, \ldots$ where $t_0, t_1, \ldots$ is a convergent sequence of time moments, and for the continuous case

$$\mu_1 = \lim_{T\to\infty} \frac{1}{T} \int_0^T \alpha(t)\mathrm{d}t \tag{1.109}$$

Similarly one can express the second moment by:

$$R(t_1, t_2) = R(s) \quad \text{for} \quad s = t_2 - t_1 \tag{1.110'}$$

where

$$R(s) = \lim_{T\to\infty} \frac{1}{T}\int_0^T x(t)x(t+s)\mathrm{d}t$$

and in the discrete case

$$R(s) = \lim_{n\to\infty} \frac{1}{n+1}\sum_{i=0}^{n} x_i x_i(s) \tag{1.110''}$$

where

$$x_i = x(t_i), \qquad x(s) = x(t_i+s), \qquad s = t_{i_2} - t_{i_1}$$

and t_i, $i = 0, 1, \ldots$ are elements of a convergent series.

The above relations can only be considered as approximations, since in a stricter sense the ergodic property of the process should be based on the mean-square limit analysis. In that case the time average of a stochastic process $x(t)$ converges in the mean-square sense to its spatial average, when $T \to \infty$.

As stated previously in statistical micromechanics, which will be discussed later, considerations will be given to the ergodic problem from a measure theoretical and probabilistic point of view by following the work of Khinchine[20], Birkhoff[21], Jancel[22] and others[23–25].

1.6 Introduction to the theory of Markov processes

(A) *Definition of a Markov process*

The theory of Markov processes is due to A. A. Markov[26], who attempted an analytical formulation of the well-known Brownian motion of statistical fluid mechanics. In earlier investigations this motion was characterized by a "transition probability" representing the probability that a particle starting from a point X is in the set $\{E\}$ at time t. Subsequently a more general theory of Markov processes was developed by A. N. Kolmogorov[27], W. Feller[28], Doob[9] and others. A comprehensive study of Markov processes is due to Dynkin[29]. The more advanced theory of Markov processes is concerned with the path of the motion $x(t)$ the mathematical description of which can then be given in terms of "transition probability measure" $\boldsymbol{P}(x(t, \omega) \in E|x(s))$, where $s, t \in T$ and $E \in \mathscr{F}$.

The fundamental condition that relates the stochastic function $x(t, \omega)$ to the measure $\boldsymbol{P}_x = \boldsymbol{P}(x(t, \omega) \in E|x(s))$ is known as the "Markov principle" in which the future is independent of the past for a known present. Thus for a known or measurable value of $x(t)$, the prediction of a subsequent motion does not depend on the character of the motion preceding the time instant t.

An important concept in the theory of Markov random processes is that of a "transition function".

Thus a function $\boldsymbol{P}(s, x, t, E)$ of the variables $s, t \in T$, $s \leqslant t$ and $x \in E \subset \mathscr{F}$ is called a *transition function of the Markov process* $x(t, \omega)$, if this function is a probability distribution on the σ-algebra $\mathscr{F}$ for fixed $s, t \in T$ and $x \in E$ and a measurable function for $x \in E$ for fixed $s, t \in T$, $E \in \mathscr{F}$ such that with probability 1:

$$\boldsymbol{P}(s, x, t, E) = \boldsymbol{P}\{x(t, \omega) \in E|x(s)\} \tag{1.111}$$

and

$$\boldsymbol{P}(s, x, s, E) = \begin{cases} 1 & \text{for} \quad x \in E \\ 0 & \text{for} \quad x \notin E \end{cases} \tag{1.112}$$

It is to be noted that a transition function $\boldsymbol{P}(s, x, t, E)$ always exists, if the topological space $(X, \mathscr{F})$ is separable and the probability distributions $\boldsymbol{P}_x = \boldsymbol{P}(x(t, \omega) \in E|x(s))$, $E \in \mathscr{F}$ are perfect measures.

For any set $s, t \in T$ and $E \in \mathscr{F}$ and almost all $x \in E$ with respect to the corresponding distributions $\boldsymbol{P}_x$ this transition function will satisfy the following relations:

$$\begin{aligned} \boldsymbol{P}(s, x, t, E) &= \sum_{y \in E} \boldsymbol{P}(s, x, t, y) \equiv \sum_{y \in E} p_{xy}(s, t) \\ p_{xy}(s, t) &= \boldsymbol{P}(s, x, t, y) \qquad\qquad (1.113) \\ p_{xy}(s, t) &= \sum_{r \in E} p_{xr}(s, u) p_{ry}(u, t), \quad s \leqslant a \leqslant t, \quad x, y \in E \end{aligned}$$

in the discrete case and

$$\boldsymbol{P}(s, x, t, E) = \int_E \boldsymbol{P}(s, x, u, \mathrm{d}y) \boldsymbol{P}(u, y, t, E), \quad s \leqslant u \leqslant t \tag{1.114}$$

in the continuous case.

The above forms represent the well-known Chapman–Kolmogorov equation. It is this functional relation that connects the stochastic theory with the theorem of "semi-groups" which will be discussed in Chapter III of this text concerned with the application of "functional analysis" in probabilistic micromechanics of solids.

Although many types of Markov processes are encountered in dealing with physical systems a strict classification of such processes is not available. In general, however, one can group Markov processes into those which have non-denumerable states and those that have a denumerable number of states. In this context perhaps a further distinction can be made as to whether these processes are:

(i) continuous in time and continuous in space
(ii) continuous in time and discrete in space
(iii) discrete in time and continuous in space
(iv) both discrete in time and space

The groups (ii) and (iii) are generally referred to as discontinuous Markov processes (see also Bharucha-Reid[30], Prohorov and Rozanov[6], Dynkin[29]). Groups (iii) and (iv) in particular are known as the class of "Markov chains". In the present formulation of probabilistic micromechanics of solids these two groups (ii) and (iv) are of particular interest and will be discussed in the later analysis.

(B) *Markov processes with a denumerable number of states* (*the Markov chain*)

(i) *The Markov property and transition probability*

Considering a stochastic process $x(t)$ and assuming that at any time $t \in Z^+$ or R^+, the transition from one "state" to another state depends not only on time t, but also on the random behaviour of a microelement, then if at time s the system is in a state "i" and goes to state "j" at a later time t, its probability which is dependent on the behaviour of the systems until time s is designated by p_{ij}. The process $x(t)$ which is characterized by such a probability is called a "Markov chain", where:

$$p_{ij}(s, t) = \boldsymbol{P}\{x(t) = j | x(s) = i\}$$
$$(i, j = 1, 2, \ldots) \qquad t, s \in T \subset Z^+ \tag{1.115}$$

These quantities are the "transition probabilities" of the Markov chain $x(t)$.

If the behaviour of the process $x(t)$ is referred to an initial time $t = t_0$, the corresponding initial probability distribution which may be experimentally accessible, is given by:

$$\overset{0}{p}_i = \boldsymbol{P}\{x(t_0) = i\}, \quad (i = 1, 2, \ldots) \tag{1.116}$$

Then

$$\boldsymbol{P}\{x(t_0) = i, x(t_1) = i_1, \ldots, x(t_n) = i_n\} = \overset{0}{p}_i p_{ii_1}(t_0, t_1) \ldots p_{i_{n-1} i_n}(t_{n-1}, t_n) \tag{1.117}$$

for any $i, i_1, \ldots, i_n$ and $t_0 \leqslant t_1 \leqslant \ldots \leqslant t_n$. The Markov chain $x(t)$ is called "homogeneous", if the transition probabilities $p_{ij}(s, t)$ depend only on the time difference, i.e.:

$$p_{ij}(s, t) = p_{ij}(t-s)$$
$$(i, j = 1, 2, \ldots), \quad s < t, \quad s, t \in T \subset Z^+ \tag{1.118}$$

(ii) *Markov time*

Some researchers employ in their analysis the concept of a "Markov time". Thus if τ denotes a random time that does not depend on the future, then for an arbitrary time t the event $\{\tau > t\}$ is determined by the behaviour of the system until t and the time τ is called a "Markov time". If E is an arbitrary event, the realization of which depends entirely on the behaviour of the system after the Markov time τ, then the system behaviour is known until τ and the probability of this event to occur coincides with the conditional probability that only the state of the system at τ is known. This can be expressed in the following form:

$$\boldsymbol{P}\{E|x(s), s \leqslant \tau\} = \boldsymbol{P}\{E|x(\tau)\} \tag{1.119}$$

In particular, for any $i_1, \ldots, i_n$ and $t_1 \leqslant t_2 \leqslant \ldots \leqslant t_n$

$$\boldsymbol{P}\{x(t_1+\tau) = i_1 \ldots x(t_n+\tau) = i_n|x(s), s \leqslant \tau\}$$
$$= p(\tau) p_{x(\tau), i_1}(\tau, t+\tau) \ldots p_{i_{n-1}, i_n}(t_{n-1}+\tau, t_n+\tau) \tag{1.120}$$

Considering now the sequence $\tau_0 \leqslant \tau_1 \leqslant \ldots \leqslant \tau_n$ in such a manner that τ_n corresponds to states where $x(\tau_n) = i_n$ is known. For instance, if τ_0 is the time instant when the system is for the first time in the state i, then the above relation (1.120) indicates that τ_1 is the instant of time when the system returns to this state, τ_2 the instant for its second return and so on. Hence for arbitrary events $E_1, E_2, \ldots, E_n$ such that every event is completely determined by the system with corresponding time interval $\tau_{n-1} \to \tau_n$ one can say that these are completely mutually independent events.

(iii) *Some remarks on the Kolmogorov equations*

The transition probabilities of a Markov chain satisfy the Kolmogorov equation (see also Prohorov and Rozanov[6]) in the following manner:

$$p_{ij}(s, t) = \sum_k p_{ik}(s, u) p_{kj}(u, t)$$

$$(i, j = 1, 2, \ldots), \qquad s \leqslant u \leqslant t \tag{1.121}$$

If $x(t)$ is a homogeneous Markov chain and $t = nh$ designates that $n = 0, 1, 2, \ldots$ and $h \geqslant 0$, then the probabilities $p_{ij}(nh)$ of a transition by means of "n-steps" are uniquely determined by the probabilities $p_{ij} = p_{ij}(h)$ of a "one-step" transition, so that:

$$p_{ij}(nh) = \sum_k p_{ik} p_{kj}[(n-1)h] = \sum_k p_{ik}[(n-1)h] p_{kj}$$

$$(i, j = 1, 2, \ldots) \tag{1.122}$$

for all $n = 1, 2, \ldots$

If the time t is continuous, and denoting specifically:

$$p_{ij}(0) = \lim_{h \to 0} p_{ij}(h) = \begin{cases} 1 & j = i \\ 0 & j \neq i \end{cases} \tag{1.123}$$

then under the assumption that the Markov chain has such a property, the probabilities are continuously differentiable for $t > 0$. Further, if the

$$\lim_{h \to 0} \frac{p_{ij}(h) - p_{ij}(0)}{h} = q_{ij}, \qquad (i, j = 1, 2, \ldots), \tag{1.124}$$

exists, where $0 \leqslant q_{ij} < \infty$ for $i \neq j$ and $q_{ij} = q_{ii} = -q_i$ for $i = j$. "q_i" is referred to as the transition density from state i. The coefficients q_{ij} are called correspondingly the *transition density from state i to j*. In terms of the abovementioned Markov time, if τ_i is the instant of time where $x(t)$ leaves the state i for the first time and

τ_{ij} the Markov time when $x(t)$ reaches j for the first time, then evidently:

$$\tau_i = \sup_{x^{(i)}(t) = i} t, \quad \tau_{ij} = \inf_{x^{(i)}(t) = j} t \tag{1.125}$$

where $x^{(i)}(t)$ is a process such that $x(0) = i$. One can therefore express the probability that the Markov chain $x(t)$ when leaving the state i and moves first to the state j, as follows:

$$\boldsymbol{P}(\tau_{ij} = \tau_i | x(0) = i) = \boldsymbol{P}(x^{(i)}(t) = j) = \frac{q_{ij}}{q_i} \qquad (i \neq j) \tag{1.126}$$

since $x^{(i)}(t)$ is a step function.

It is to be noted that the above transition densities always satisfy the inequality:

$$\sum_{i \neq j} q_{ij} \leqslant q_i, \quad (i = 1, 2, \ldots) \tag{1.127}$$

Under these conditions the transition probabilities $p_{ij}(t)$ satisfy the so-called "backward Kolmogorov equations". Under certain restrictions such as the boundedness of q_{ij} for example the system of the "forward Kolmogorov equations" will also hold, i.e.:

$$p'_{ij}(t) = \sum q_{kj} p_{ik}(t), \quad (i, j = 1, 2, \ldots) \tag{1.128}$$

In the case of a homogeneous Markov chain and when t is continuous, then $q_i \to \infty$. Thus when reaching the state i, the system leaves it instantaneously with probability one or:

$$\boldsymbol{P}\{\tau_i = 0 | x(0) = i\} = 1 \tag{1.129}$$

Hence for any arbitrary small time interval Δt after $t = 0$, and if Δi denotes the time spent in the state i during Δt one can write:

$$\boldsymbol{P}\left\{\lim_{\Delta t \to 0} \frac{\Delta i}{\Delta t} = 1 | x(0) = i\right\} = 1 \tag{1.130}$$

Thus the state i is "stable" if $q_i < \infty$. Once the system is in the stable state i it remains there with probability one. The Markov chain $x(t)$ is a stable one, if with probability one the system passes during an arbitrary finite period only a finite number of times from one state to another.

In conclusion it should be noted that the brief presentation of Markov processes and, in particular, the considerations of a Markov chain will be discussed again in subsequent chapters. The Markov principle is of particular importance in the present theory since it permits to relate the experimentally accessible probability measures to those derived from analytical considerations.

II. Random Theory of Deformation

2.1 Introduction

In order to formulate the "random theory of deformation" according to the approach taken in probabilistic micromechanics, it will be necessary to discuss first the fundamental notions on which this theory is based. Since the theory deviates in many respects from the conventional continuum theory, it may be instructive to give a comparison between both these approaches. This will be done in Section 3 of this chapter, which also attempts to briefly outline the "statistical micromechanics" formulation. A comprehensive study of the latter is however beyond the scope of this monograph and must be deferred to a later publication.

It has been mentioned earlier, that essentially two groups of structured solids will be discussed in this text, i.e. polycrystalline solids and fibrous structures. Hence the subsequent treatment of the deformation kinematics will only deal with these groups. The theory indicated in the last section of this chapter is more general in that it is considered to be independent of the structured medium under consideration. Moreover, a "general theory of stochastic deformation" will be introduced in Chapter III after considerations have been given to the use of functional analysis and the adoption of an operator formalism, which is essential in the formulation of a general deformation theory. In the subsequent analysis majuscules will denote quantities in the "undeformed state" of the structured solid, whilst their corresponding interpretations in the "deformed state" will be written in minuscules. Greek letters to the left of a parameter will indicate the individual microelement and capital Latin superscripts on the left refer to mesoscopic quantities.

It should be noted that whilst the deformation parameters in the case of a purely elastic response of the medium can be interpreted as "random variables" for a proper deformation process or a steady-state deformation such variables represent a "stochastic process" and are recognized as such by having the time t in their basic argument. Finally, all subsequent considerations will refer to the "deformation space" only. The relation of the latter to the corresponding "stress space" will be treated later, when the concept of a material operator is discussed on the basis of functional analysis (Chapter III).

2.2 Basic concepts of the random theory of deformation

The new random theory of deformation of structured solids is based on the following four fundamental concepts[31–34].

(i) The first concept concerns the use of three measuring scales, the smallest of which refers to a structural or microelement of the material which is designated by its volume ${}^{\alpha}v$ and its surface ${}^{\alpha}s$. The presuperscript is used for its identification in the existing microstructure. For simplification of the analysis one can, as a first approximation of the internal contribution to the material response assume, that the constitutive relations of a microelement (internal) are describable by continuum laws, whilst the stress and displacement fields are described either in terms of random variables or stochastic processes depending on the loading and the medium under investigation. All parameters concerned with this region are prefixed by "micro". Next an intermediary scale is introduced referred to as a "mesodomain" that contains a statistical ensemble of microelements α ($\alpha = 1, 2, \ldots, N$; N large). This domain is interpreted to be much smaller than the "macroscopic domain" of the entire material body, but is much larger than the domain of a microelement. Indeed, it is viewed to be similar to a Gibbsian representative ensemble of statistical mechanics. From a micromechanics point of view such a domain is analytically and experimentally seen as a physical region of the macroscopic material body within which the statistics (i.e. probability density, distribution function, etc.) of any random variable or stochastic process can be assumed to be spatially homogeneous. This at once takes into account the nature of the particular medium when forming the mesodomains as well as the specific boundary value problem to be considered in a given application.

In the case of polycrystalline solids the microelement is a single crystal or grain. The mesodomain is then a statistical ensemble of these elements, whose probabilistic quantities whether they are kinematic or material characteristics are independent of position within this particular domain. For fibrous systems, however, a microelement does not consist of a single fibre, but rather of the unsupported segment between two bonding areas and one half of each bond area between fibres. Again the mesodomain is chosen so that it contains a statistical ensemble of such elements and for which all probabilistic quantities including those that are experimentally accessible, are independent of position within the mesodomain. This does not exclude however the possible effect of the fibre orientation within the mesodomain. For the purpose of illustrating this first concept, the mesodomains of some structured solids are indicated in Fig. 10 and Fig. 11 below. Figure 10 shows the three measuring scales generally adopted in probabilistic micromechanics.

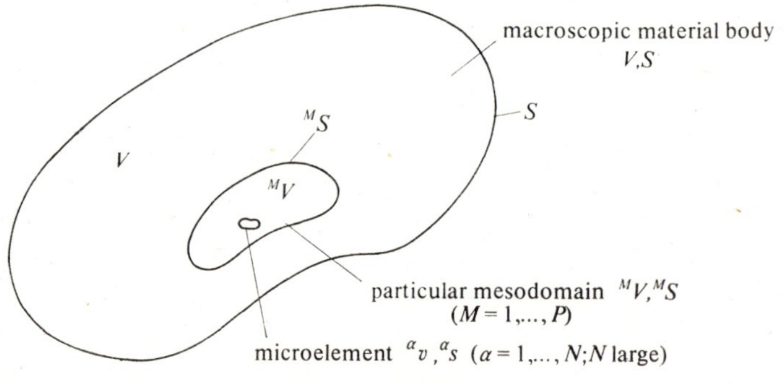

Fig. 10. Macro, meso and micro domains.

It is assumed in micromechanics in accordance with the first concept above, that the macroscopic material body is formed by a denumerable number of non-intersecting mesodomains such that:

$$^{M}V \cap {}^{L}V = \emptyset, \qquad L \neq M$$

and

$$V = \bigcup_{M=1}^{P} {}^{M}V, \qquad V \gg {}^{M}V \gg {}^{\alpha}v, \qquad (M = 1, \dots, P) \tag{2.1}$$

For a polycrystalline solid Fig. 11 (a) shows a "closed packed" arrangement of the microelements, whilst cluster arrangements and isolated elements embedded in a matrix of different physical properties are classified as two-phase structures Fig. 11 (b). The model of a soil microstructure is indicated in Fig. 11 (c) in which the two kaoline platelets kept together by an electrostatic potential can be taken as a microelement. The microelement of a two-dimensional fibrous network and a corresponding mesodomain are shown in Fig. 11 (d), which also indicates a "theoretical scanning line" and a "scanning area" the meaning of which will be fully discussed later in the text.

(ii) The second concept or postulate in the stochastic theory is that all microscopic field quantities in the formulation are considered to be either random variables or stochastic functions of these "primitive random variables". These quantities will be more precisely defined in the subsequent analysis.

(iii) The third concept of fundamental importance is concerned with the application of the "stress principle". It is evident that the Cauchy stress principle cannot be applied in the formulation of interaction effects between microelements. It can be used only to simplify the analysis and as a first approximation to the response of single elements, when the latter are considered as homogeneous. In the boundary zone between two microdomains a "generalized surface force" must be introduced. Such a force is derivable from the "surface potential" between two adjacent microelements. In the case of polycrystalline solids a detailed discussion on this surface force has been given in reference[35] in which Bollman's theory[36] and Goux's[37] grain boundary model have been used.

For fibrous systems the "bonding potential" that exists between overlapping fibres is considered for the determination of this generalized force.

(iv) The fourth fundamental concept in this theory and which is used in the formulation of the response behaviour of a structured

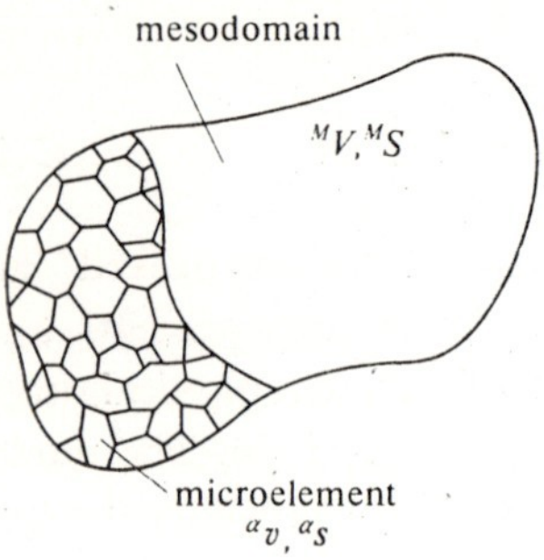

(a) Closed packed microelements (metal)

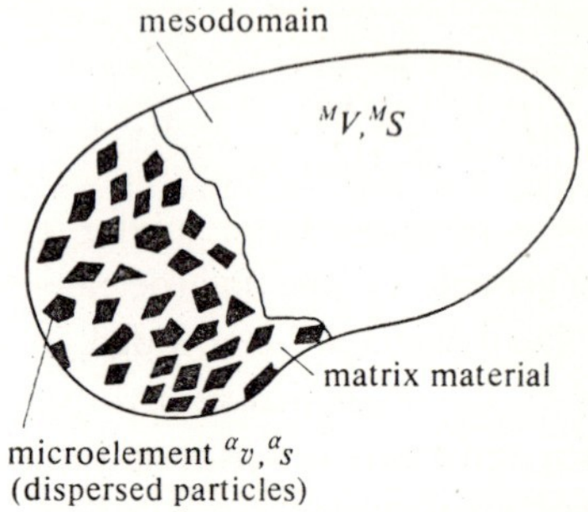

(b) Two-phase structure (ceramic-metal)

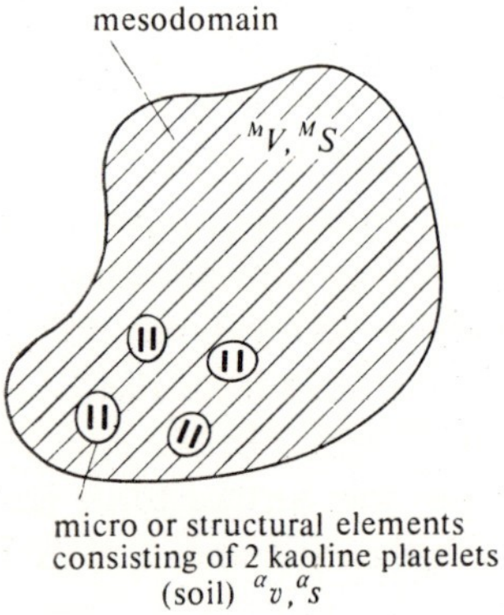

(c) Soil

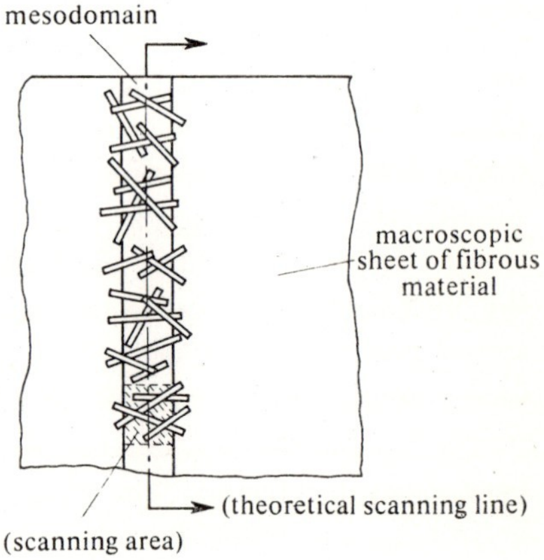

(d) Two-dimensional model of a fibrous network

Fig. 11. Mesodomains of some microstructures.

medium, is the existence of a "material functional" or "material operator ${}^{M}\mathscr{M}$" that is characteristic for a specific material and which contains geometrical and thermomechanical parameters or functions of such parameters. This functional connects the stresses and deformations in a structured medium and hence replaces the conventional constitutive relations of continuum mechanics. It can be written as a characteristic energy functional in terms of the important field variables for a polycrystalline solid, for instance, as follows:

$$ {}^{M}\mathscr{M} = {}^{M}\mathscr{M}\{\boldsymbol{E}, \boldsymbol{a}, \varrho_d, \Psi, {}^{\alpha}\boldsymbol{O}, {}^{\alpha}\boldsymbol{\sigma}, {}^{\alpha}v; \theta, t\} \tag{2.2} $$

where the full meaning of these parameters will be discussed in later sections of this text. One can also write this material operator in the case of a steady-state deformation in an operational form briefly as:

$$ {}^{M}\mathscr{M} = {}^{M}\mathscr{M}\{A, B, F; \theta, t\} \tag{2.3} $$

in which A, B represent stochastic integro-differential operators characterizing the response of an ensemble of microelements including interaction effects within a particular mesodomain. The function F in this expression is the distribution function of the significant field variables involved in the description of the stochastic deformation process, θ the temperature and t the time.

It is evident from the above four fundamental concepts of probabilistic micromechanics that the approach taken in the present theory differs considerably from the conventional one of continuum mechanics. To further outline this difference a discussion is presented in the next section of this chapter which is concerned with a comparison between probabilistic micromechanics and the continuum mechanics approach.

2.3 Comparison between continuum mechanics and probabilistic micromechanics

(A) *Fundamental concepts*

In order to describe the corresponding concepts of continuum and probabilistic micromechanics of solids two tables have been compiled that serve to indicate the differences in the analysis. Thus, Table II below contains some fundamental aspects of both formulations.

TABLE II

Some fundamental aspects of continuum theory and probabilistic micromechanics

Classical continuum mechanics	Probabilistic micromechanics
Material: continuous system	Material: discrete system
Individual particle (grain) identified by geometric point	Individual particle = microelement of volume ${}^{\alpha}v$, boundary ${}^{\alpha}s$
Deformation kinematics of "continuous system"	Deformation kinematics (probabilistic) of "discrete system" (interaction effects included)
Deterministic deformation theory	Random theory of deformation including interaction effects
Cauchy stress principle Continuous stresses and deformations	Generalized surface force (internal) Discontinuous deformations or strains
Balance equations for macroscopic continuum	Balance equations for mesodomains
Constitutive theory	Material operator ${}^{M}\mathscr{M}$ for mesodomain (operational formalism)
Macroscopic response relations	Governing eqn. for structured solid, macroscopic response relations

It is seen that from the onset of the analysis and according to the first concept of the probabilistic theory the structured solid is considered as a "discrete system" in which individual microelements are given a finite volume and boundary. Hence the kinematics of deformation are studied for discrete media and are distinctly different from those of a continuum. Furthermore, since the second concept introduces a probabilistic description of the kinematic and thermomechanical properties of a structured solid a "random theory of deformation" is used, which differs from the conventional deterministic formulation of continuum mechanics. An even more significant difference between the two approaches arises from the third fundamental concept that uses a "generalized surface force" to account for interaction effects between individual microelements. Thus, this concept admits two possible deformations, i.e. an internal one and an interfacial deformation. Whilst for simplicity the internal deformation can be ascribed to the action of a Cauchy stress within an individual microelement, if the latter is considered as a continuum, this cannot be done in the case of the interfacial deformation. This type of deformation must be interpreted in terms of a generalized surface force caused by the existing interaction potential between the surfaces of two adjacent microelements. The adoption of a generalized force based on a form described by Jancel[22] and the concept of generalized functions by Gel'fand and Vilenkin[16] (see Chapter I) as well as the consideration of the surface potential discussed from a statistical mechanics point of view by Yvon[38] leads then in a probabilistic formulation to the random theory of deformation of a structured solid including interaction effects.

Perhaps the most significant departure from the conventional formulation of continuum mechanics is the introduction of the fourth fundamental concept or the use of a "material functional". It is apparent from the above statements that "constitutive relations" for a structured solid must differ greatly from those of the classical theory. In fact, in many cases it is not possible to obtain a simple constitutive relation in the usual sense. However, by the use of a "material operator" an equivalent functional type of relationship between stress and deformation or stress and strain can be accomplished. Thus using a "material functional" and the relevant distribution functions of the field quantities involved in the formulation, the response behaviour of the structured medium can be derived first for a mesodomain. Such relations can then be extended to obtain the macroscopic response of the material. Since in probabilistic micromechanics the material is regarded as a discrete system which obeys certain statistical laws within a mesodomain, it is apparent that an analogy between the classical statistical mechanics and statistical micromechanics will have to be established. For this purpose Table III below has been compiled to indicate the corresponding steps to be taken in the formulation. In this context it should be noted that a comprehensive discussion on statistical micromechanics is beyond the scope of this monograph and hence only a brief review can be given subsequently.

In the classical description a representative point system occupying the phase-space Γ with generalized coordinates q_i, momenta p_i and time t, is considered. Then another space, the so-called μ-space in the sense of Ehrenfest[39] is introduced in order to obtain a "fine grained" and "coarse grained" density in these spaces, respectively. The coarse graining of the Γ-space is necessary so that an "evolution criterion" of the physical process can be obtained. However, since

TABLE III

Comparison between statistical formulations

Classical statistical mechanics	Statistical micromechanics
(1) Representative point system generalized coordinates: q_i, p_i, t Motion in Γ-space	(1) Representative state vector $\mathbf{v}(t)$ State space X Probabilistic function space
(2) Phase-space, Γ-, μ-space, coarse graining	(2) Measure space X with appropriate topology
(3) Densities: ϱ_Γ, ϱ_μ	(3) Abstract dynamical system $[X, \mathscr{F}, \mathscr{P}]$
(4) Ergodic theorem Perturbation theory	(4) Markov theory: quadruple $[X, \mathscr{F}, \mathscr{P}, Q]$
(5) Randomness assumption Master equation	(5) Governing equation for structured medium
(6) Macroscopic relations	(6) Macroscopic relations

the problem of describing macroscopic irreversibility from a "microscopic reversible" phenomenon cannot be done in a mathematical rigorous manner using deterministic concepts only, usually a "randomness assumption" for the process is made and a "master equation" for the medium is formulated which renders a macroscopic relation possible. This procedure is well known for gases and fluids in classical statistical mechanics and is discussed amongst others by Kac[40], Uhlenbeck[41] and van Kampen[42].

(B) *Statistical formulation in micromechanics*

The procedure adopted in classical statistical mechanics as indicated in the first column of Table III is of a different sequence in the second column of Table III which outlines the "statistical micromechanics". Thus in this approach the mechanical states of the microelements are considered by a set of r-dimensional state vectors ${}^{\alpha}\mathbf{v}(t): {}^{\alpha}v_i(t)$, $\alpha = 1, \ldots, N$; $i = 1, \ldots, r$.

These vectors are, in general, stochastic functions of random geometric and mechanical parameters involved in the formulation of the response behaviour of a structured medium. Hence the "state space" to which these vectors belong is identified in this theory by a "probabilistic function space X". Since a structured solid, according to the present theory, is regarded as a collection of a finite number of mesodomains each of which contains a statistical ensemble of microelements, two other sets can be formed such that:

$$ {}^{M}X = \{{}^{\alpha}\mathbf{v}(t);\ \alpha = 1, \ldots, N\} \tag{2.4}$$

and for the entire macroscopic body:

$$ \mathscr{X} = \{{}^{M}X;\ M = 1, \ldots, P\} \tag{2.5}$$

where P is the number of denumerable, non-intersecting mesodomains. Although it is possible to introduce a σ-algebra "$\mathscr{F}$" of Borel sets in the state space "X" together with an appropriate probability measure "$\mathscr{P}$" to form an "abstract dynamical system $[X, \mathscr{F}, \mathscr{P}]$" as indicated in line 3 of Table III, it is rather difficult to topologize this general state space. Hence for the convenience of the subsequent analysis "subspaces" of this state space will be introduced by means of which kinematic and other mechanical parameters involved in the formulation of the deformational behaviour of a structured medium can be distinguished. The triplet $[X, \mathscr{F}, \mathscr{P}]$ representing an abstract dynamical system and thus the motion of a structured medium, will contain all possible thermomechanical characteristics of a particular medium. However, if only the purely kinematic quantities involved in the motion of the medium are considered, then one can introduce a finite dimensional "subspace Ω" of the probabilistic function space X. On the basis of the axiomatic definition of probability, the space Ω can be considered as a sample space formed by the positive outcomes ${}^{\alpha}\omega$ of an experiment or trial for an individual microelement α ($\alpha = 1, \ldots, N$), where ${}^{\alpha}\boldsymbol{\omega} = {}^{\alpha}\omega_i$ ($i = 1, \ldots, r$) identifies the number of basic kinematic parameters of interest in a given problem. Thus, if ${}^{\alpha}\boldsymbol{\omega} \in \Omega$, it can be associated with the change of kinematic quantities related to the "undeformed" and "deformed" configuration of the structured medium. However, due to experimental constraints, it is only possible to specify ${}^{\alpha}\omega$ to within a certain range ${}^{\ni}\omega$, ${}^{\ni}\omega+\Delta^{\ni}\omega$, where the superscript $\ni$ designates this particular range and $\Delta^{\ni}\omega$ denotes the accuracy with which observations are possible. It is evident that this range can be represented by an open set or sphere or event ${}^{\ni}E$ such that:

$$ {}^{\ni}E = \{{}^{\ni}\omega_i < {}^{\alpha}\omega_i < {}^{\ni}\omega_i + \Delta^{\ni}\omega_i;\ i = 1, \ldots, r\} \tag{2.6} $$

in which ${}^{\ni}E \subset \Omega$ and $\ni \in Z^+$ (see also Sneddon[14]). Hence one can define a class $\mathscr{F}$ of these spheres ${}^{\ni}E \in \mathscr{F}$ (see Section 1.3), where $\mathscr{F}$ forms a σ-algebra in the subspace Ω. In order to complete the considerations in the general kinematic subspace in terms of the triplet $[\Omega, \mathscr{F}, \mathscr{P}]$, it is necessary to introduce a probability measure $\mathscr{P}\{{}^{\alpha}\omega \in {}^{\ni}E\}$, which in accordance with the definition given in expression (2.6) and Section 1.3, can be written as follows:

$$ \mathscr{P}\{{}^{\alpha}\omega \in {}^{\ni}E\} = \mathscr{P}\{{}^{\ni}\omega_i < {}^{\alpha}\omega_i < {}^{\ni}\omega_i + \Delta^{\ni}\omega_i\} \tag{2.7} $$

On the assumption that each element ${}^{\alpha}\omega_i \in {}^{\ni}E$ and has equal probability to be in the set ${}^{\ni}E \in \mathscr{F}$, the probability measure in the discrete case can be conveniently expressed by:

$$ \mathscr{P}\{{}^{\alpha}\omega \in {}^{\ni}E\} \equiv \mathscr{P}\{{}^{\alpha}\omega : {}^{\alpha}\omega_i = {}^{\ni}\eta_i\} \tag{2.8} $$

where ${}^{\ni}\eta_i$ designates a measurable value of ${}^{\alpha}\omega_i$ in the set ${}^{\ni}E$. It is of interest to note that this step is related to the concept of mesodomains in a structured solid. Thus considering N microelements ($\alpha = 1, \ldots, N$; N large) within a particular mesodomain, it can be visualized that instead of following the stochastic trajectories of the basic kinematic parameters of different microelements for N-experiments, each performed under identical external conditions, one can consider only one experiment in which the time evolution of the parameters of the statistical ensemble of microelements ($\alpha = 1, \ldots, N$) is noted. This implies the assumption of the law of large numbers of the theory of probability and that one can represent the discrete system of microelements by separate "trials or experiments". Whilst in the study of the deformation kinematics within the reversible range of a structured solid the basic quantities may be regarded as simple vector random variables, a more general view must be taken for the description of the deformation process as a whole. Hence the kinematic quantities are then defined as stochastic processes (see Section 1.5) or a class of random variables $\{{}^{\alpha}\omega(t); t \in T\}$, indexed by t belonging to the set T. Thus in view of the above probability measure (2.8), a discrete probability distribution function pertaining to the stochastic process ${}^{\alpha}\omega(t)$ can be written as:

$$F({}^{\ni}\boldsymbol{\eta}, t) = \mathscr{P}\{{}^{\alpha}\omega(t) \leqslant {}^{\ni}\boldsymbol{\eta}\} \tag{2.9}$$

Continuing the discussion on statistical micromechanics according to Table III, in line (4) the Markovian assumption for the description of a general deformation process is made. From this point of view the "conditional probability" in terms of equation (1.5) can be expressed as follows:

$$\begin{aligned} \boldsymbol{P}(t_1) &= \mathscr{P}\{\omega(t_2) = {}^{\ni}\boldsymbol{\eta} | \omega(t_1) = {}^{\text{я}}\boldsymbol{\eta}\} \\ &= \frac{\mathscr{P}\{\omega(t_2) = {}^{\ni}\boldsymbol{\eta}, \omega(t_1) = {}^{\text{я}}\boldsymbol{\eta}\}}{\mathscr{P}\{\omega(t_1) = {}^{\text{я}}\boldsymbol{\eta}\}} \end{aligned} \tag{2.10}$$

in which $\ni, \text{я} \in Z^+$, $t_1 < t_2$ and where $\boldsymbol{P}$ indicates the conditional probability of the stochastic deformation being in the kinematic state ${}^{\ni}\boldsymbol{\eta}$ at time t_2, when it was in the state ${}^{\text{я}}\boldsymbol{\eta}$ at time t_1.

As already indicated in Section 1.6 of the preceding chapter the transition probability $\boldsymbol{P}$ under the restriction that the deformation process is approximated in terms of a homogeneous Markov chain, is governed by the well-known Chapman–Kolmogorov equation (1.113) as follows:

$$\boldsymbol{P}(t_2) = \boldsymbol{P}(t_1)\boldsymbol{P}(t_2 - t_1) \tag{2.11}$$

Letting now $t_1 = t$, $t_2 - t_1 = \Delta t$, then in the limit when $\Delta t \to 0$ the time rate of change of the transition probability in terms of the Kolmogorov differential equation (1.128) becomes:

$$\frac{d}{dt}\boldsymbol{P}(t) = \boldsymbol{Q}\boldsymbol{P}(t) \tag{2.12}$$

where $\boldsymbol{Q}$ is in general a time-dependent transition probability matrix given by:

$$\boldsymbol{Q} = \lim_{\Delta t \to 0} \frac{\{\boldsymbol{P}(t) - \boldsymbol{I}\}}{\Delta t} \tag{2.13}$$

in which $\boldsymbol{I}$ is the identity matrix. A more detailed discussion of this formulation for the case when $\boldsymbol{Q}$ is time-independent and the possibility of its assessment from experimental observations has been given in an earlier publication[43]. The relation (2.12) is of utmost significance in the probabilistic micromechanics of structured media, in particular, for the development of a general stochastic deformation theory as discussed in Chapter III of this monograph. Thus it will be shown subsequently that such "transition matrices" can be written for the subspaces of stress and deformation, respectively, and that they are related by the characteristic material operator. In order to conclude the comparison between the classical and the statistical micromechanics approach it should be mentioned that earlier work considered the formulation of a so-called "Master equation" of a structured solid[44–46] and that from a topological point of view the representation of the abstract dynamical system by inclusion of the transition matrix takes the form of a quadruple $[X, \mathscr{F}, \mathscr{P}, Q]$ in the general state space (column 5, Table III).

2.4 Deformation kinematics

(A) *Introduction*

In order to formulate a general stochastic deformation theory as will be shown in the following chapter of this monograph, it is necessary to deal with the kinematics of deformation of a structured solid first. Although the general theory is applicable to any type of structured media, the deformation kinematics of the two classes of materials, i.e. polycrystalline solids and fibrous systems will be considered here. In the following presentation majuscules will denote quantities in the "undeformed state" of the solid, whilst their corresponding interpretations in the "deformed state" will be written in minuscules. Greek letters to the left of a parameter will indicate the individual microelement or correspondingly a boundary between them. Capital Latin superscripts on the left refer generally to mesoscopic quantities. The relevant deformation parameters in the case of the reversible or purely elastic response of the structured solid are considered to be "primitive" random variables and in the more general deformation process to be represented by stochastic functions of these variables. The latter are recognized as such by having time t in their basic arguments. The subsequent analysis will refer to the "deformation-space" only. The relation of this space to its conjugate "stress-space" will be treated later, when the role of the "material functional" or "material operator" in the context of the general stochastic deformation theory will be discussed on the basis of functional analysis in micromechanics (Chapter III).

(B) *Deformation kinematics of polycrystalline solids*

The kinematic parameters introduced in this paragraph will be only those of specific use in the formulation of the stochastic deformation theory. Hence velocities and accelerations as well as other quantities that may be involved in a more general study of the deformation kinematics are not included here and therefore no attempt of completeness will be made at this stage in the presentation. Using the concept of mesodomains of a macroscopic material sample the "undeformed" and "deformed" configurations of a particular domain as well as the employed internal and external reference frames are shown schematically in Fig. 12 below. Thus the position vector to

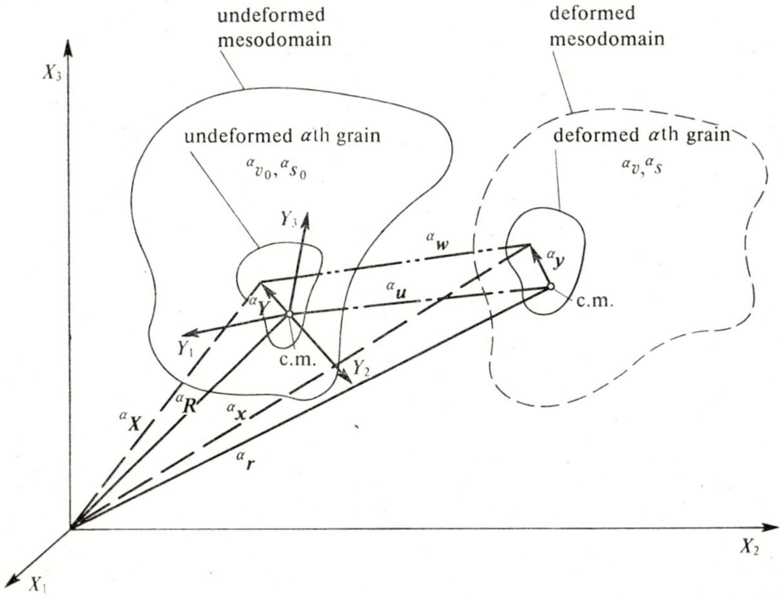

Fig. 12. Kinematics of single crystals or microelements α, of a particular mesodomain ${}^{M}V$, ${}^{M}S \supset {}^{\alpha}v$, ${}^{\alpha}s$ ($\alpha = 1, \ldots, M$).

any arbitrary point within the αth crystal relative to a fixed external frame (X_1, X_2, X_3) is given by:

$$^{\alpha}\boldsymbol{X} = {}^{\alpha}\boldsymbol{O}^{\alpha}\boldsymbol{Y} + {}^{\alpha}\boldsymbol{R}$$

or (2.14)

$$^{\alpha}X_I = {}^{\alpha}O_{I\Delta}{}^{\alpha}Y_{\Delta} + {}^{\alpha}R_I \qquad (I, \Delta = 1, 2, 3)$$

in which $^{\alpha}\boldsymbol{O}$ is the matrix of the direction cosines between the crystallographic axes of the grain α and the fixed Cartesian frame and $^{\alpha}\boldsymbol{R}$ is the position vector to the centre of mass (C.M.) of the crystal. During and at the end of a random deformation the position of this point in the crystal will be:

$$^{\alpha}\boldsymbol{x} = {}^{\alpha}\boldsymbol{o}^{\alpha}\boldsymbol{y} + {}^{\alpha}\boldsymbol{r}$$

or

$$^{\alpha}x_i = {}^{\alpha}o_{i\delta}{}^{\alpha}y_{\delta} + {}^{\alpha}r_i \qquad (i, \delta = 1, 2, 3) \tag{2.15}$$

As mentioned previously in order to include grain boundary effects in the deformational behaviour of a polycrystalline solid, it is necessary to distinguish between arbitrary points interior to the crystal as indicated above and points belonging to the surface of the crystal, i.e. to the grain boundary. Thus considering a point on the "internal surface" (see also Fig. 13) of the αth grain, relations (2.14) and (2.15) become:

$$^{\alpha}\boldsymbol{G} = {}^{\alpha}\boldsymbol{O}^{\alpha}\boldsymbol{H} + {}^{\alpha}\boldsymbol{R}, \qquad {}^{\alpha}\boldsymbol{g} = {}^{\alpha}\boldsymbol{o}^{\alpha}\boldsymbol{h} + {}^{\alpha}\boldsymbol{r} \tag{2.16}$$

where $^{\alpha}\boldsymbol{G}$, $^{\alpha}\boldsymbol{H}$ and $^{\alpha}\boldsymbol{g}$, $^{\alpha}\boldsymbol{h}$ are the position vectors of this surface point in the undeformed and deformed state, respectively. Thus using the above relations it is evident that the random displacement for the centre of mass of a grain (Fig. 13) will be:

$$^{\alpha}\boldsymbol{\Omega} = {}^{\alpha}\boldsymbol{r} - {}^{\alpha}\boldsymbol{R} \tag{2.17}$$

and for an arbitrary point inside the grain:

$$^{\alpha}\boldsymbol{w} = {}^{\alpha}\boldsymbol{x} - {}^{\alpha}\boldsymbol{X} = {}^{\alpha}\boldsymbol{o}^{\alpha}\boldsymbol{y} - {}^{\alpha}\boldsymbol{O}^{\alpha}\boldsymbol{Y} + {}^{\alpha}\boldsymbol{\Omega} \tag{2.18}$$

$$^{\alpha}\boldsymbol{w} - {}^{\alpha}\boldsymbol{\Omega} = {}^{\alpha}\boldsymbol{y} - {}^{\alpha}\boldsymbol{Y} \tag{2.19}$$

Analogously, the displacement at the surface of the αth grain can be written as:

$$^{\alpha}\boldsymbol{\mu} = {}^{\alpha}\boldsymbol{g} - {}^{\alpha}\boldsymbol{G} = {}^{\alpha}\boldsymbol{o}^{\alpha}\boldsymbol{h} - {}^{\alpha}\boldsymbol{O}^{\alpha}\boldsymbol{H} + {}^{\alpha}\boldsymbol{\Omega} \tag{2.20}$$

The above geometric together with other physical parameters can be regarded as the basic quantities involved in the response behaviour of a polycrystalline solid. In order to assess the contribution due to the existence of grain boundaries as shown schematically in Fig. 13 the simple kinematic model will be extended to include this effect.

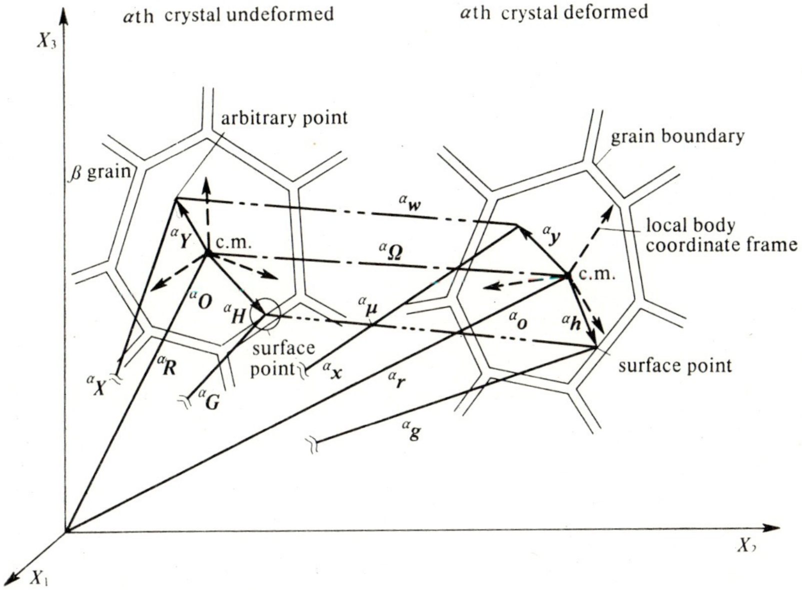

Fig. 13. Idealized model for the deformation kinematics of the αth grain or microelement.

For this purpose as shown in previous publications[35,47] a probabilistic "surface molecular coincidence lattice" model will be used. This model is based on the geometrical theory of coincidence lattices by Bollman[36] and Goux's[37] grain boundary studies.

The schematics of a typical undeformed "coincidence cell" in the grain surface is indicated below in Fig. 14 (a, b).

In Bollman's theory grain boundaries are defined analytically in terms of coincidence lattices obtained from the interpenetration of two neighbouring grains and where the coincidence lattice points form equivalent groups to the lattice points of the crystals. On the other hand, Goux considers the grain boundary to have a fairly disordered or amorphous structure that separates the two adjacent grain surfaces (Fig. 14(a)). However the latter model admits the existence of a certain distance between lattice points of two adjacent grains. Hence it is evident that the model used in the present theory may be regarded

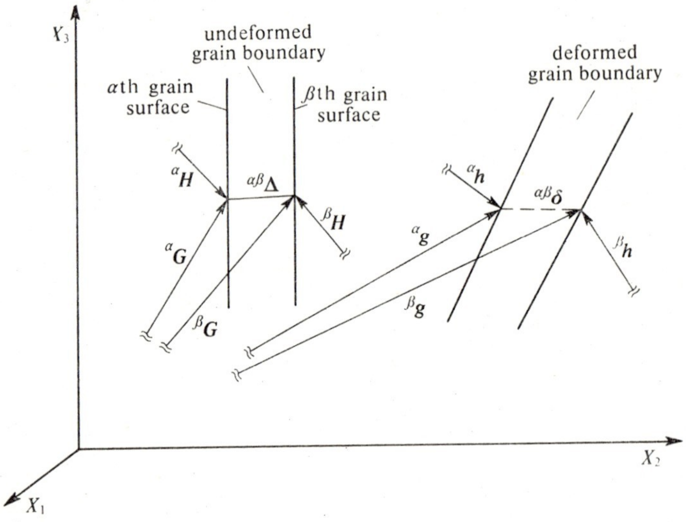

(a) Grain surface displacement

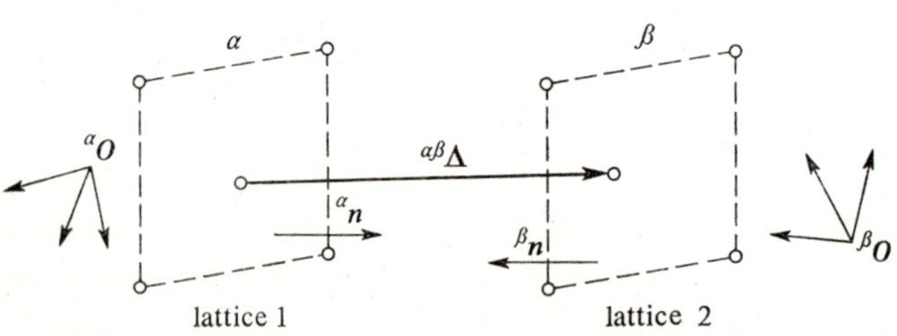

(b) Coincidence cell model area $\Delta^q S$

Fig. 14. Grain boundary kinematics of two adjacent grains α, β.

as a combination of the two representations mentioned above. Thus the "distance vector between two coincidence points" of two adjacent grain surfaces (α, β) is considered here to be the basic kinematic parameter in the assessment of the grain boundary deformation. This kinematic variable is referred to as "relative displacement $\boldsymbol{d}$" of two adjacent surfaces caused by the application of external loads. The "undeformed" distance between the crystal lattice of each adjacent grain, which is unaware of the presence of a "local strain field" induced in the grain boundary or which may also be referred to as the "calculation zone" of the inherent grain boundary energy may be expressed in accordance with the previously introduced coordinate system as follows:

$$^{\alpha\beta}\boldsymbol{\Delta} = {}^{\beta}\boldsymbol{G} - {}^{\alpha}\boldsymbol{G} = {}^{\beta}\boldsymbol{O}^{\beta}\boldsymbol{H} - {}^{\alpha}\boldsymbol{O}^{\alpha}\boldsymbol{H} + {}^{\beta}\boldsymbol{R} - {}^{\alpha}\boldsymbol{R} \tag{2.21}$$

This distance vector is indicated in Fig. 14(b) together with the corresponding orientation tensors $^{\alpha}\boldsymbol{O}, {}^{\beta}\boldsymbol{O}$. In the deformed state this distance may be written as:

$$^{\alpha\beta}\boldsymbol{\delta} = {}^{\beta}\boldsymbol{g} - {}^{\alpha}\boldsymbol{g} = {}^{\beta}\boldsymbol{o}^{\beta}\boldsymbol{h} - {}^{\alpha}\boldsymbol{o}^{\alpha}\boldsymbol{h} + {}^{\beta}\boldsymbol{r} - {}^{\alpha}\boldsymbol{r} \tag{2.22}$$

so that the stochastic "relative displacement" becomes:

$$^{\alpha\beta}\boldsymbol{d} = {}^{\alpha\beta}\boldsymbol{\delta} - {}^{\alpha\beta}\boldsymbol{\Delta} = {}^{\beta}\boldsymbol{\mu} - {}^{\alpha}\boldsymbol{\mu} \tag{2.23}$$

in which relation (2.20) has been used.

The above kinematic parameters and some other material characteristics which will be discussed in Chapter IV, can be established from their corresponding distribution functions. The information concerning these distribution functions is obtained from experimental observations that will be described in Chapter V of this monograph.

Thus for instance the distribution of the crystallographic orientations $(^{\alpha}\boldsymbol{O}, {}^{\alpha}\boldsymbol{o})$ in the undeformed and deformed state of the material, can be established by X-ray diffraction techniques [48–52]. The distribution of the grain size can be readily obtained from micrographic studies. Similarly, other significant parameters contained in the material operator $^{M}\mathscr{M}$ (eqn. (2.2)) which is required in the formulation of the response behaviour of a structured solid, are equally experimentally accessible. From the beginning of the studies in probabilistic micromechanics, one of the main hypotheses has been the Gaussian nature of these distribution functions and that they are statistically homogeneous, non-isotropic and applicable throughout a particular mesodomain, although they may vary from meso to mesodomain. This permitted from the onset and for a simplification of the analysis to employ correlation theory (see for instance Yaglom[8]) at least for steady-state deformation so that the distribution functions can be completely specified by their first and second statistical moments. In this context a characteristic correlation parameter for the case of the simple model considered here (Figs. 12, 13) can be chosen as the distance vector between the centres of mass of two adjacent elements or crystals, such that

$$\boldsymbol{\lambda} = {}^{\beta}\boldsymbol{R} - {}^{\alpha}\boldsymbol{R} \tag{2.24}$$

Thus the mathematical expectation and correlation functions for the abovementioned orientations and grain sizes will be given by:

$$\begin{aligned} E\{{}^{\alpha}\boldsymbol{O}\} &= \langle \boldsymbol{O} \rangle_M, \quad B_O(\boldsymbol{\lambda}) = \langle \boldsymbol{O}({}^{\alpha}\boldsymbol{R})\,\boldsymbol{O}({}^{\alpha}\boldsymbol{R}+\boldsymbol{\lambda})\rangle_M \\ E\{{}^{\alpha}v\} &= \langle v \rangle_M, \quad B_v(\boldsymbol{\lambda}) = \langle v({}^{\alpha}\boldsymbol{R})\,v({}^{\alpha}\boldsymbol{R}+\boldsymbol{\lambda})\rangle_M \end{aligned} \tag{2.25}$$

where "B" with the corresponding suffix represents the correlation function (see Section 1.4), of the relevant quantity. In an analogous manner "microdeformations" will be considered in paragraph (iv) of this section.

Returning to the schematics of the undeformed "coincidence cell" model (Fig. 14(b)) then, in general, the surface between any two crystals will take up such a position that the crystals exhibit more or less optimum matching. Thus on the assumption that lattice 1 is fixed and lattice 2 is changing, i.e. undergoing a translation and rotation, the latter will be translated in such a manner that at least one point coincides with a point on the surface of lattice 1. This point is referred to as "coincidence site". Due to the inherent periodicity of the two lattices assuming that they are "ideal", i.e. free of any defects, a finite number of such points will exist. These points form then from a purely geometrical point of view another lattice termed "coincidence lattice". Hence Bollman's theory leads to a grain boundary topology in terms of two idealized crystals, that form an interpenetrating mathematical translation lattice, which depends of course on the "misfit angle θ", between the given crystal lattices. It is for the latter reason that the theory uses a so-called "O-lattice" for the case of low misfit angles and an "O_2-lattice" for high misfit angles. A more detailed discussion on the grain boundary structure, the model used in the present theory, and its contribution to the overall response behaviour of a polycrystalline solid, follows in Section 4.2 of Chapter IV.

As a measure of the "coincidence cell" lattice one may consider the surface of its "unit cells", which in turn also depends on the lattice vector ($\boldsymbol{a}$, relation (2.2)) of the specific crystal structure under consideration as well as on the "relative orientation". Since in the present analysis the orientations of the adjacent crystals ${}^{\alpha}\boldsymbol{O}, {}^{\beta}\boldsymbol{O}$ (Fig. 14 (b)) are taken as random quantities, the coincidence areas denoted by ${}^{\alpha}\Delta s^q = {}^{\beta}\Delta s^q$ $(q = 1, \ldots, p; p =$ number of coincidence cells) will be random functions of both ${}^{\alpha}\boldsymbol{O}$ and ${}^{\beta}\boldsymbol{O}$. Whilst this will be discussed later in the text, it may be stated that in general the size and shape of the boundary coincidence cells are dependent on the distribution of the misfit angle $\mathscr{P}(\theta)$ occurring in the grain boundary region.

From a kinematics point of view designating the "undeformed" distance vector by ${}^{\alpha\beta}\boldsymbol{\Delta}$ (2.21) and that in the "deformed state" by ${}^{\alpha\beta}\boldsymbol{\delta}$, the change in these vectors becomes:

$$ {}^{\alpha\beta}\hat{\boldsymbol{d}} = {}^{\alpha\beta}\hat{\boldsymbol{\delta}} - {}^{\alpha\beta}\hat{\boldsymbol{\Delta}} \tag{2.26} $$

in which the "$\hat{}$" sign indicates here the discrete character of these parameters. The undeformed distance vector ${}^{\alpha\beta}\hat{\boldsymbol{\Delta}}$ may be regarded as an instantaneous equilibrium distance between matching points of the undeformed grain surfaces α, β. The basic kinematic parameter throughout this analysis is therefore the relative vector ${}^{\alpha\beta}\hat{\boldsymbol{d}}$. Again the basic significance of this vector in dealing with the inclusion of interaction effects in the response behaviour of a structured solid will be given in more detail in later sections of this monograph.

(C) *Deformation kinematics of fibrous systems*

The treatment of the deformation kinematics of fibrous materials follows closely that of crystalline solids discussed briefly in the preceding paragraph. Only the basic kinematic relations pertaining to a microelement of the structure will be discussed below. The effects of these considerations on the overall response behaviour of fibrous systems will be studied in Section 4.3 of Chapter IV.

A microelement of a fibrous material as already indicated in Fig. 5 is defined in the present theory to consist of the "unsupported segment" of a single fibre between two "bonded areas" and one half of each bonding area ${}^{\alpha\beta}A$. In this context it should be mentioned that a distinction must be made between the optically observable total bonding area and the effective or actual one ${}^{b}A$, which is brought about by the actual number of bonds that exist between two overlapping fibres. A more detailed study of such bonding areas follows later in Section 4.3 on the basis of a three-dimensional model of the bond structure. The undeformed configuration of such an element is indicated in Fig. 15 below, where ${}^{\alpha}\boldsymbol{R} \approx {}^{\beta}\boldsymbol{R}$ are the position vectors to the centre of mass (C.M.) of a bonding area ${}^{\alpha\beta}A$ or ${}^{b}A$, respectively. This figure also shows a "theoretical scanning line M–M" the meaning of which will be discussed later in Section 4.3. It suffices to say, that such a line is required for the evaluation of the experimental observations of the surface deformations, which occur in a two-dimensional fibrous network. It is convenient for the study of kinematics and completely analogous to the treatment of crystalline solids to use a body frame ${}^{\alpha}\boldsymbol{Y}$ and external frame $\boldsymbol{X}$, where the orientation of the former with respect to the $\boldsymbol{X}$-frame is given by ${}^{\alpha}\boldsymbol{O} = \{\cos(\boldsymbol{e} \cdot \boldsymbol{n})\}$ (Fig. 15). The kinematics of a microelement in this case is however complicated by the presence of bonding areas. The latter can be studied again in an analogous manner to the previously discussed coincidence lattice theory by employing for the entire bonding zone between the αth and βth fibres "coincidence cells", whose equilibrium distance is denoted by ${}^{\alpha\beta}\boldsymbol{\Delta}^{q}$ $(q = 1, \ldots, p)$ and where p is the number of actual hydrogen bonds corresponding to the "matching points" in the overlapping fibres. For the case of cellulose fibres for example these points are indicated in Fig. 16 (b) by 2 and 2′, respectively, following the standard notation of the atomic arrangement forming the intramolecular hydrogen bond (see also Chapter IV). In general, the actual number of lattice points and the size of the "repeating or unit cells" contained in the actual bonding area depends largely on the type of fibrous structure and the size of the former. Thus the size of a unit cell used for the coincidence lattice model within the bonding area for natural cellulose has been found to be of the order of 10.3×8.35 Å.

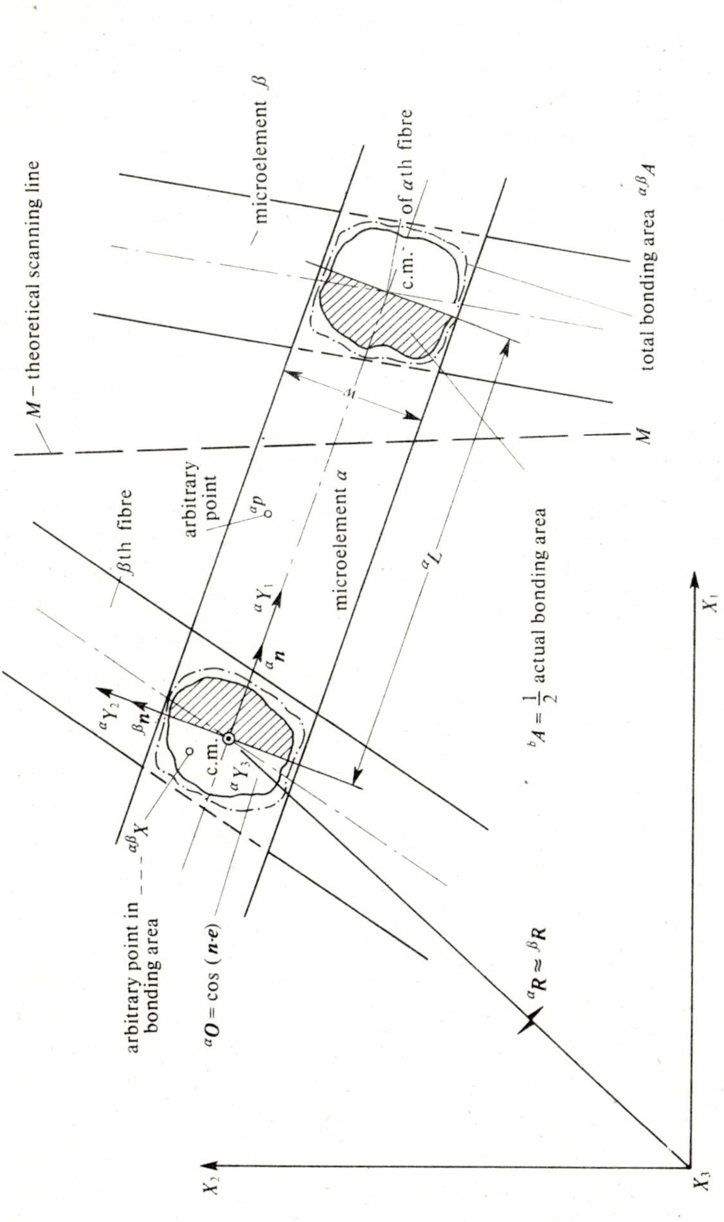

Fig. 15. Microelement of a fibrous system with bonding areas $^{\alpha\beta}A$ (internal reference frame Y, external fixed reference frame X) (undeformed state).

It is seen with reference to Fig. 16 (a) that the motion of an arbitrary point ${}^{\alpha}p$ within the fibre segment to its deformed position ${}^{\alpha}p'$ can be expressed by:

$$\begin{aligned} {}^{\alpha}\Omega_i: &\quad {}^{\alpha}\boldsymbol{\Omega} = {}^{\alpha}\boldsymbol{r} - {}^{\alpha}\boldsymbol{R} \\ {}^{\alpha}w_i: &\quad {}^{\alpha}\boldsymbol{w} = {}^{\alpha}\boldsymbol{x} - {}^{\alpha}\boldsymbol{X} \end{aligned} \tag{2.27}$$

so that

$$\begin{aligned} {}^{f}u_I: &\quad {}^{f}\boldsymbol{u} = {}^{\alpha}\boldsymbol{y} - {}^{\alpha}\boldsymbol{Y} \\ {}^{f}u_i: &\quad {}^{f}\boldsymbol{u} = {}^{\alpha}\boldsymbol{x} - {}^{\alpha}\boldsymbol{X} - {}^{\alpha}\boldsymbol{\Omega} \end{aligned}$$

and

$${}^{f}u_i = {}^{\alpha}\Lambda_{Ii}{}^{f}u_I \tag{2.28}$$

in which the meaning of the symbols is the same as discussed previously and where ${}^{\alpha}\Lambda_{Ii}$ represents the transformation matrix from the internal to the external reference frame indicated by the subscripts I, i, respectively. The deformation vectors as shown in the fibre will be discussed more fully in Chapter IV (Section 4.3). Using the notation introduced in Section B for the relative displacement vector, the latter in the case of a fibrous microelement can be written as:

$${}^{\alpha\beta}\boldsymbol{d}^q = {}^{\alpha\beta}\boldsymbol{\delta}^q - {}^{\alpha\beta}\boldsymbol{\Delta}^q \tag{2.29}$$

so that a complete analogy between the polycrystalline solids and fibrous structures can be established, if the interfacial relations are considered on the basis of a coincidence cell theory.

(D) *General deformation kinematics*

The basic kinematic relations for the two groups of structured solids considered here have been given in the preceding paragraphs. It is readily noticed that in contrast to continuum mechanics, where the motion of a material point is described by one set of vectors, i.e. for example $\boldsymbol{u}(\boldsymbol{X}; t)$ that are functions of the position vector $\boldsymbol{X}$ of the centre of mass in the undeformed configuration and time t, the stochastic deformation kinematics employs two sets of vectors ${}^{\alpha}\boldsymbol{\omega}({}^{\alpha}\boldsymbol{X}; t)$ and ${}^{\alpha}\boldsymbol{\mu}({}^{\alpha}\boldsymbol{X}; t)$. The random vectors ${}^{\alpha}\boldsymbol{\omega}$ and ${}^{\alpha}\boldsymbol{\mu}$ represent the motion of an arbitrary point within and that of a surface point of a structural element. It is further possible to use equivalently for the

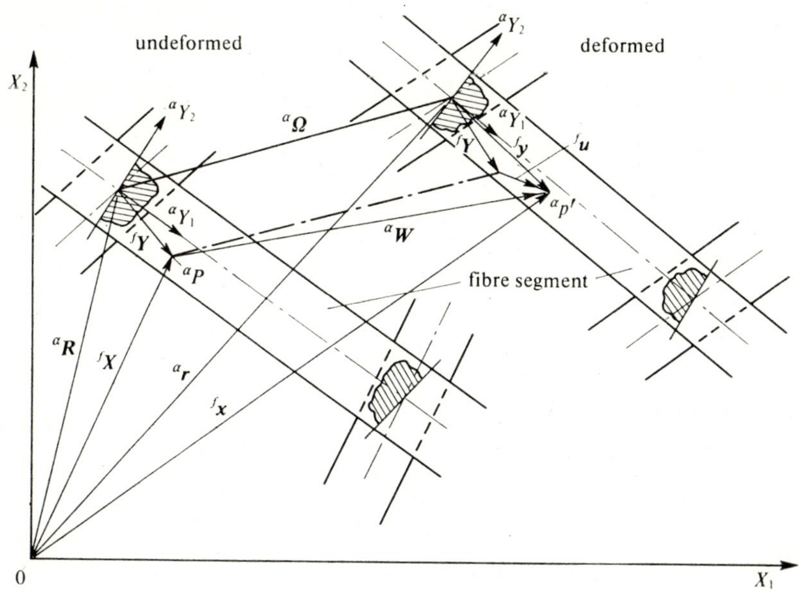

(a) Kinematics of a fibre segment

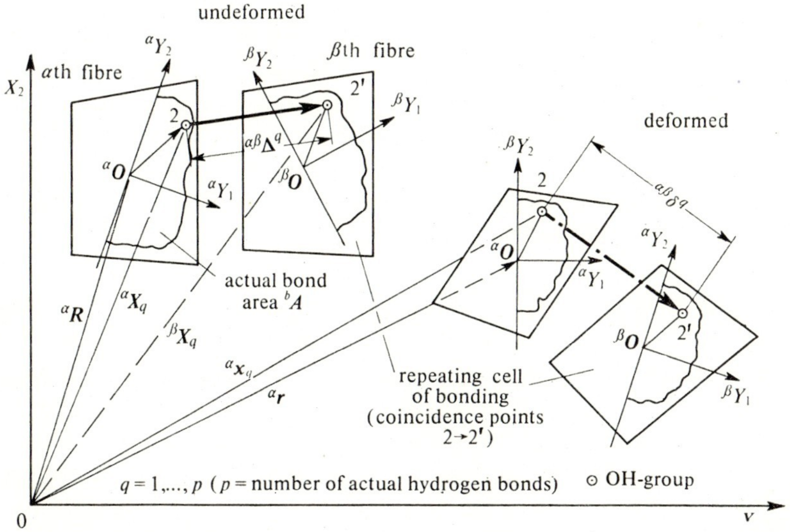

(b) Kinematics of bond area (coincidence cell model matching points 2-2′ of the fibre-fibre interface)

Fig. 16. Kinematics of "fibrous microelement".

inclusion of the interfacial kinematics a "relative distance vector ${}^{\alpha\beta}\boldsymbol{d}$" for both types of materials. The corresponding relations were given in (2.18), (2.27) and (2.23), (2.29), respectively. Thus, if ${}^{\alpha}\boldsymbol{u}({}^{\alpha}\boldsymbol{X}; t)$ denotes for instance the total deformation of a microelement, it can be expressed as a random vector function by:

$$ {}^{\alpha}\boldsymbol{u}({}^{\alpha}\boldsymbol{X}; t) = f[{}^{\alpha}\boldsymbol{w}({}^{\alpha}\boldsymbol{X}; t), {}^{\alpha}\boldsymbol{\mu}({}^{\alpha}\boldsymbol{X}; t)] \tag{2.30} $$

or by using $\boldsymbol{d}$, also in the form of:

$$ {}^{\alpha}\boldsymbol{u}({}^{\alpha}\boldsymbol{X}; t) = g[{}^{\alpha}\boldsymbol{w}, {}^{\alpha\beta}\boldsymbol{d}] \tag{2.31} $$

The form of the function "f" for the elastic behaviour of polycrystalline solids has been investigated from a system theory point of view. It will be discussed further, when dealing with the application of the stochastic deformation theory and also in the analysis of the mechanical relaxation behaviour of structured solids. It was shown that ${}^{\alpha}\boldsymbol{u}$ can be considered as a linear function of ${}^{\alpha}\boldsymbol{w}$ and ${}^{\alpha\beta}\boldsymbol{d}$ such that:

$$ {}^{\alpha}\boldsymbol{u} = L_1{}^{\alpha}\boldsymbol{w} + L_2{}^{\alpha\beta}\boldsymbol{d} \tag{2.32} $$

in which L_1 and L_2 are two filtering operators discussed subsequently in Chapter IV. In deriving the above relation (2.32) it was implicitly assumed that the vectors ${}^{\alpha}\boldsymbol{w}$, ${}^{\alpha\beta}\boldsymbol{d}$ are sufficiently continuous functions of $({}^{\alpha}\boldsymbol{X}; t)$ within a compact support, e.g. that the deformation fields are locally continuous. On this assumption, it is possible to define a strain measure in the classical sense by using a "microdeformation gradient" defined for the internal deformations only, by:

$$ {}^{\alpha}\boldsymbol{F} = \frac{\partial^{\alpha}\boldsymbol{y}}{\partial^{\alpha}\boldsymbol{Y}}, \quad {}^{\alpha}\boldsymbol{f} = \frac{\partial^{\alpha}\boldsymbol{Y}}{\partial^{\alpha}\boldsymbol{y}} \tag{2.33} $$

from which a Lagrangian or Eulerian strain can be expressed as follows:

$$ {}^{\alpha}\boldsymbol{E} = \frac{1}{2}({}^{\alpha}\boldsymbol{F}^{c}\cdot{}^{\alpha}\boldsymbol{F} - \boldsymbol{\delta}), \quad {}^{\alpha}\boldsymbol{\varepsilon} = \frac{1}{2}(\boldsymbol{\delta} - {}^{\alpha}\boldsymbol{f}^{c}\cdot{}^{\alpha}\boldsymbol{f}) \tag{2.34} $$

in which $\boldsymbol{\delta}$ is the Kronecker delta and the superscript "c" denotes the dyadic conjugate. Similarly a gradient for the interfacial zone between elements, if such a zone is considered for simplicity as a continuum can be written in terms of the surface coordinate as follows:

$$ {}^{\alpha\beta}\boldsymbol{F} = \frac{\partial^{\alpha}\boldsymbol{h}}{\partial^{\alpha}\boldsymbol{H}}, \quad {}^{\alpha\beta}\boldsymbol{f} = \frac{\partial^{\alpha}\boldsymbol{H}}{\partial^{\alpha}\boldsymbol{h}} \tag{2.35} $$

so that the corresponding strains become:

$$ {}^{\alpha\beta}\boldsymbol{E} = \frac{1}{2}({}^{\alpha\beta}\boldsymbol{F}^{c}\cdot{}^{\alpha\beta}\boldsymbol{F} - \boldsymbol{\delta}), \quad {}^{\alpha\beta}\boldsymbol{\varepsilon} = \frac{1}{2}(\boldsymbol{\delta} - {}^{\alpha\beta}\boldsymbol{f}^{c}\cdot{}^{\alpha\beta}\boldsymbol{f}) \tag{2.36} $$

It is equally possible to express the above intinitesimal strains in terms of the vectors ${}^{\alpha}\boldsymbol{w}$ and ${}^{\alpha\beta}\boldsymbol{d}$ such that:

$$ {}^{\alpha}\boldsymbol{\varepsilon} = \nabla^{\alpha}\boldsymbol{w} \qquad \text{(a)} $$

and (2.37)

$$ {}^{\alpha\beta}\boldsymbol{\varepsilon} = \nabla^{\alpha\beta}\boldsymbol{d} \qquad \text{(b)} $$

where in (2.37 a) the gradient operator is taken with respect to the internal body frame and in (2.37 b) with respect to the surface coordinate frame attached to the interfacial surface.

The above forms are based on the assumption that the material within and at the boundary between two adjacent microelements can be regarded as a continuum. However, in real structured materials due to the presence of defects and a rather disordered medium in the boundary zone, the random vectors ${}^{\alpha}\boldsymbol{w}$ and ${}^{\alpha\beta}\boldsymbol{d}$ will not be continuous functions of ${}^{\alpha}\boldsymbol{X}$ and thus the deformation fields will be discontinuous or discrete. Hence, in order to construct a more general theory of deformation as discussed subsequently, it is necessary to introduce "generalized" quantities such as generalized deformations and strains, which are continuous functions of ${}^{\alpha}\boldsymbol{X}$. This approach does not change the actual physical characteristics of the structured medium under consideration, but serves to define conveniently the kinematic parameters involved in the analysis. In this manner one can use a "generalized microstrain ${}^{\alpha}\boldsymbol{\varepsilon}$" of a structural element as a function of $\nabla^{\alpha}\boldsymbol{u}$ such that:

$$ {}^{\alpha}\boldsymbol{\varepsilon} = f(\nabla^{\alpha}\boldsymbol{u}) \tag{2.38} $$

In particular, for the polycrystalline solid, if the local strains (2.37 a, b) exist, one can write:

$$ {}^{\alpha}\boldsymbol{\varepsilon} = f({}^{\alpha}\boldsymbol{\varepsilon}, {}^{\alpha\beta}\boldsymbol{\varepsilon}) \tag{2.39} $$

and by using the form of (2.32), then:

$$^{\alpha}\boldsymbol{\varepsilon} = L_1 \nabla^{\alpha} \boldsymbol{w} + L_2 \nabla^{\alpha\beta} \boldsymbol{d} \tag{2.40}$$

in which L_1 and L_2 are filtering operators as shown in Section 4.4 of Chapter IV. A similar interpretation of generalized quantities can be given for fibrous structures, although it must be noted that in that case the body coordinate and surface coordinate frames are the same, whilst in crystalline solids they are not.

In conclusion of this section concerned with the deformation kinematics of the two types of interacting systems, it has been shown that explicit expressions for the overall deformations of a micro-element can be given. However, these expressions are restricted by the underlying simplified assumptions that have been made and do not represent the more general case. Furthermore, whilst continuity of the random vectors $^{\alpha}\boldsymbol{w}$, $^{\alpha}\boldsymbol{\mu}$ or $\boldsymbol{d}$ can be assumed under certain restrictions this will not represent the general case. In particular, these quantities in actual structured solids are discrete quantities. Thus in order to investigate the essential properties of the above deformation fields and to formulate a more general theory of stochastic deformations it becomes necessary to introduce functional analytic concepts in the analysis. It is the main purpose of the following chapter to consider the aspects of functional analysis and the operational formalism required in the development of such a general theory in the probabilistic micromechanics of structured solids.

III. Functional Analysis in Micromechanics

3.1 Introduction

In the preceding chapters the basic concepts of probabilistic micromechanics and the deformation kinematics of two classes of structured solids have been discussed. It has been pointed out that in order to formulate a more general theory of deformation for discrete materials, it is necessary to employ functional analysis. The latter is also required in statistical micromechanics and for the use of the operational formalism in which the general deformation theory is presented.

Thus, in dealing with the deformational behaviour of discrete systems, which, to a varying degree exhibit interaction effects that influence the overall response behaviour, an "abstract dynamical system" $[X, \mathscr{F}, \mathscr{P}]$ can be introduced, in which X represents a general probabilistic function space, $\mathscr{F}$ the σ-algebra of Borel sets in this space and $\mathscr{P}$ an appropriate probability measure. As in classical statistical mechanics (see Section 2.3 of Chapter II), one has first to specify admissible vector fields in the physical domains so that the possible states of either individual microelements or an ensemble of them within a particular mesodomain of the macroscopic material body can be represented in a "state space". By identifying this space with the "probabilistic function space", the notion of continuity can be applied to the abstract dynamical systems. However, in order to achieve a formulation of the general deformation process and more specifically its evolution with time, it becomes necessary to give the general function space and its associated "subspaces" an appropriate topological structure and a suitable probability measure.

It is well known from statistical mechanics of continuous media (see for instance Birkhoff, Bona and Kampé de Fériet[53]), when such media are considered as conservative systems that they can be represented by a set of hyperbolic differential equations that are associated with groups of transformations T_t on the function space X with the domain $-\infty < t < \infty$. However, in the case of discrete systems and materials considered in this text, they will be represented by a set of differential equations that induce only semi-groups $\{T_t\}$ of transformations defined for $t \geqslant 0$. Hence it will be necessary to study certain theorems of the theory of semi-groups that apply to particular topological vector spaces such as Fréchet spaces.

As mentioned previously in the presentation of a general theory of stochastic deformations, an operational formalism will be used and this process will be considered to be approximated in general by a Markov process. This approach leads to the Chapman–Kolmogorov relation, which connects the stochastic theory with that of semi-groups. Since in this theory the material functional or material operator replaces the conventional constitutive relations, it will be necessary to discuss this operator from a functional analytic point of view. It is evident from the above statements that the functional analytic aspects in probabilistic micromechanics require a brief review of certain theorems of topology and measure theory as given below.

3.2 Topological considerations

This section is aimed at sketching the concepts involved in the application of functional analysis to the micromechanics of solids. First a brief account on "mappings" or transformations of sets into other sets is given, then certain definitions and theorems related to the structure of topological spaces required in the subsequent analysis are briefly discussed. The concept of sets mentioned already in Section 1.3 of Chapter I lies at the foundation of functional analysis.

(A) *Mappings*

By a mapping of one set X into another set Y is meant a rule that associates a unique element of Y with each element of X. Such a mapping is symbolically written as:

$$f: X \to Y \tag{3.1}$$

in which X is called the "domain" of the mapping f, or $D(f)$; $f(x)$ indicates the element of Y corresponding to the element x of X and is called the "image" of x under the mapping f.

Consider the set $M \subset X$ and $f: X \to Y$, then the image set of M is given by:

$$f(M) = \{y \in Y; y = f(x) \text{ for } x \in M\} \tag{3.2}$$

When $M = X$, the image set is referred to as the range $R(f)$ of the mapping. The following definitions of mappings are standard:

(a) f is "one-to-one", if for each $x, y \in X; f(x) = f(y) \Rightarrow x = y$; this mapping is called "injective".

(b) f is "onto" if $f(X) = Y$; then this mapping is "surjective".

(c) If f is both injective and surjective, the mapping is "bijective".

In general mappings are known in mathematics under various names such as functions, transformations, operators and operations. Another definition of mapping frequently used is that of "inverse images". Thus

(d) $f^{-1}(y)$ defines the inverse image of y under the mapping f^{-1} if

$$f^{-1}(y) = \{x \in X; f(x) = y\} \tag{3.3}$$

It is seen from this definition that $f^{-1}(y) \neq \emptyset$, if and only if $y \in f(X)$. Considering a subset $N \subset X$ then the above definition can be extended to:

$$\text{(e)} \quad f^{-1}(N) = \{x \in X;\ f(x) \in N\} = \bigcup_{y \in N} f^{-1}(y) \tag{3.4}$$

defining the "inverse image set". Thus for any set X one can write an "identity mapping" as:

$$I_x(x) = x, \qquad x \in X \tag{3.5}$$

Further, considering $f: X \to Y$ and $g: Y \to Z$ one can also define a "product mapping" by:

$$(g \circ f)(x) = g(f(x)), \qquad x \in X \tag{3.6}$$

However the commutative law does not hold for products of mappings, but the associative one does so that:

$$h \circ (g \circ f) = (h \circ g) \circ f \tag{3.7}$$

It follows from (3.5) and (3.6) that the mapping $g: Y \to X$ is the inverse of $f: X \to Y$. In the case when the latter is bijective, then

$$f^{-1} \circ f = I_x, \quad f \circ f^{-1} = I_Y \tag{3.8}$$

Hence for a bijective mapping it is possible to define an inverse. If f is not a bijective, one can still construct an inverse mapping in a restrictive sense (see for instance Sneddon[14]).

(B) *Topological spaces*

Due to the need of constructing a proper topology for the function spaces utilized in probabilistic micromechanics a brief discussion on topological spaces is given below. For a more extensive treatment the reader is referred to the texts of Kolmogorov and Fomin[54], Dieudonné[55], Taylor[56], Blumenthal and Getoor[57], Simmons[58], Sneddon[14] and others[59,60]. The study of topological spaces entails the provision of a set with a structure so that considerations can be given to the problem of convergence and continuity. In order to achieve a more general concept of continuity required in the application of functional analysis to the deformation of discrete materials, one could employ the notion of the neighbourhood of an element of the set. However, it is more convenient for the present analysis to arrive at a definition of continuity in terms of "open sets".

It is well known from the theory of real variables that a set X of real numbers is "open", if and only if for every element $x \in X$ there exists a number $\alpha(x)$ such that the open interval $(x-\alpha, x+\alpha)$ is contained in X. It can be readily verified that an open interval is an open set and that such sets on R^+ (the real line) have the property that any union or finite intersection of the open sets is open. On the assumption that in certain sets there exist "subsets" with the above properties, it is possible to use these sets for the definition of continuity and the corresponding spaces are then "topological spaces".

Thus a topology on X is defined to be a family $\mathscr{T}$ of subsets of X with the following properties:

A.1: The union of sets in any subfamily of $\mathscr{T}$ belongs to $\mathscr{T}$.

A.2: The intersection of the sets in any finite subfamily of $\mathscr{T}$ is contained in $\mathscr{T}$. (3.9)

A.3: The null set $\emptyset$ and the whole set X are both in $\mathscr{T}$.

In terms of a family $\mathscr{T}$ of "open sets" the above theorems can be stated as follows:

A.1: Any union of open sets is open.
A.2: Any finite intersection of open sets is open. (3.10)
A.3: $\emptyset$ and X are open sets.

If the above axioms (3.10) are satisfied, then the class $\mathscr{T}$ is called a topology for X whilst the set X together with its topology $\mathscr{T}$ is called a topological space $[X, \mathscr{T}]$. It should be noted that there is always a trivial topology in which the null set $\emptyset$ and the set X are the only open sets. On the other hand there is also a "discrete topology" in which every subset of X is an open set.

It is perhaps of interest to note that a mapping f of a topological space $[X_1, \mathscr{T}_1]$ into another space $[X_2, \mathscr{T}_2]$ is called continuous, if for every set $X \in \mathscr{T}_2$ the inverse image $f^{-1}(X)$ is open, that is if $f^{-1}(X) \in \mathscr{T}_1$. A one-to-one mapping f of X_1 to X_2 is a "homeomorphism", if f and f^{-1} are continuous.

A topological space is called a "Hausdorff space", if for every pair of distinct points x, y of the space there are disjoint open sets X_1, X_2 with $x \in X_1$ and $y \in X_2$. It is significant that a Hausdorff space is defined in terms of "separation properties". Such properties are discussed in some detail in the previously mentioned references. A "metric space" $[X, d]$ is a non-empty set of elements with a non-negative distance function $d(x, y)$ defined for all pairs of elements $x, y \in X$ satisfying the conditions below for all points $x, y, z \in X$:

$$
\begin{aligned}
&\text{A.1:} \quad d(x, y) \geqslant 0 \\
&\text{A.2:} \quad d(x, y) = 0 \quad \text{iff} \quad x = y \\
&\text{A.3:} \quad d(x, y) = d(y, x) \\
&\text{A.4:} \quad d(x, y) \leqslant d(x, z) + d(y, z)
\end{aligned}
\tag{3.11}
$$

The above distance function is called a "metric". For instance the set of all real variables $(x_1, x_2, \dots, x_n, y_1, y_2, \dots, y_n)$ with the distance function:

$$d(x, y) = \Big\{\sum_i (x_i - y_i)^2\Big\}^{1/2}$$

form a metric space. If the open sets are generated by the "base $\{y; d(x, y) < \alpha\}$" for all $x \in X$, $\alpha > 0$, the metric space is a Hausdorff space.

Another significant topology is the "product topology" $\mathscr{T}_1 \times \mathscr{T}_2$. Thus given two topological spaces $[X_1, \mathscr{T}_1]$, $[X_2, \mathscr{T}_2]$ a topology defined on the product space is then given by:

$$\begin{aligned} \mathscr{T}_1 \times \mathscr{T}_2 &= \{O_1 \times O_2; O_1 \in \mathscr{T}_1, O_2 \in \mathscr{T}_2\} \\ O_1 \times O_2 &= \{(x, y); x \in O_1, y \in O_2\} \end{aligned} \tag{3.12}$$

It is seen from (3.11) that any set on which a metric can be defined gives a metric space. However, if X is a metric space, there is not necessarily a unique metric defined on it. Hence different metrics will induce different topologies. If however two distinct metrics define the same topology on X they are called "equivalent". In general, not every topological space is such that its topology is derivable from a metric. Whenever this is possible the space is called "metrizable". If X is a metric space with the metric d and if "a" is a fixed point of X, "ε" a positive number, then the sets of points:

$$K(a, \varepsilon) = \{x;\ x \in X;\ d(x, a) < \varepsilon\} \tag{3.13}$$

defines an "open sphere" with the centre a and radius ε.

Considering now a subset $N \subset X$ to be open for every point $a \in N$, there will be a positive number $\varepsilon = \varepsilon(a)$ such that $K(a, \varepsilon) \subset N$ whereby the axioms of open sets (3.10) are satisfied. Generally the distance of a point "a" from the set A in the metric space $[X, d]$ can be defined therefore, by:

$$d(a, A) = \inf_{x \in A} d(a, x) \tag{3.14}$$

where the infimum always exists since $d(a, A)$ is always positive and only equal to zero, if and only if $a \in \bar{A}$.

(C) *Topological vector spaces*

Consider a set $\{X\}$ of elements $x, y, \ldots$ which is a continuous group under addition. Assuming that the multiplication with any real (or complex) number $\alpha, \beta, \ldots$; i.e. that αx is well defined and that $x \in X$, $\alpha x \in X$, then if the conditions:

$$\begin{aligned} &\text{(i) } \alpha(x+y) = \alpha x + \alpha y \\ &\text{(ii) } (\alpha+\beta)x = \alpha x + \beta x \\ &\text{(iii) } (\alpha\beta)x = \alpha(\beta x) \\ &\text{(iv) } 1 \cdot x = x \end{aligned} \tag{3.15}$$

for all elements $x, y \in X$ and the numbers α, β are satisfied, X is called a "linear vector space". A non-empty subset N of a linear space X is called a "linear manifold or vector subspace" of X, if:

$$\begin{aligned} &\text{(i) } x+y \in N \text{ whenever } x \text{ and } y \in N \\ &\text{(ii) } \alpha x \in N \text{ for } x \in N \end{aligned} \tag{3.16}$$

So far as the mapping of one vector space onto another is concerned, a one-to-one correspondence $f: X_1 \to X_2$ is a linear mapping, if and only if:

$$\begin{aligned} f(x+y) &= f(x)+f(y) \\ f(\alpha x) &= \alpha f(x) \end{aligned} \tag{3.17}$$

If $x, y \in X$, then one defines a "segment" by:

$$\overrightarrow{xy} = \{a \in X;\ a = \alpha x+(1-\alpha)y;\ \alpha \in R^+,\ 0 \leqslant \alpha \leqslant 1\} \tag{3.18}$$

A subset A of a vector space X is "convex", if and only if:

$$x, y \in A \Rightarrow \overrightarrow{xy} \subset A \tag{3.19}$$

The vector space X is said to be "normed", if there is a non-negative function or norm, denoted by $||\cdot||$ defined on X such that:

$$\begin{aligned} &\text{(i) } ||x|| \geqslant 0; \quad ||x|| = 0 \text{ if and only if } x = 0 \\ &\text{(ii) } ||x+y|| \leqslant ||x||+||y|| \\ &\text{(iii) } ||\alpha x|| = |\alpha| \cdot ||x|| \end{aligned} \tag{3.20}$$

in which $x, y \in X$ and α a real (complex) number. The function $||\cdot||$ which assigns to each vector $x \in X$ the real number $||x||$ is a norm of X, if and only if the conditions (3.20) are met for all $x, y \in X$. A sequence $x_n \in X$; $n = 1, 2, \ldots$ is said to be a "Cauchy sequence", if

$$||x_n - x_m|| \to 0 \quad \text{as } m, n \to \infty \tag{3.21}$$

The normed vector space X is a "Banach space", if it is complete, that is, if every Cauchy sequence (3.21) has a limit $x \in X$ such that $||x_n - x|| \to 0$ whenever $n \to \infty$.

Designating the set of sequences $x = (x_j)$, where x_j $(j = 1, 2, \ldots)$ are real (or complex) numbers by l^p with the norm:

$$||x|| = \Big[\sum_{j=1}^{\infty} |x_j|^p\Big]^{1/p} < \infty \quad \text{for a fixed } p > 0 \tag{3.22}$$

and defining the addition and multiplication as follows: $x+y = (x_i + y_i)$, $\alpha x = (\alpha x_i)$ for $x = (x_i)$, $y = (y_i)$, then for $1 \leqslant p < \infty$ the l^p space is a Banach space with the norm (3.22). In a similar manner the space l^∞ will be a Banach space of bounded sequences $x = (x_j)$, $j = 1, 2, \dots$, with the norm:

$$||x|| = \sup_j |x_j| < \infty \tag{3.23}$$

Of special interest here is the L^p space. If ω denotes a σ-finite measure on the Borel field of subsets $\Omega \subset X$, then the set of all "ω-measurable functions $f(x)$" on the space X with the measure ω which are integrable and for which

$$||f|| = \left[\int |f(x)|^p \mathrm{d}\,\omega(x)\right]^{1/p} < \infty, \quad 1 \leqslant p < \infty \tag{3.24}$$

forms an L^p-space with the norm (3.24) and the pointwise addition and multiplication by a number.

A special type of Banach space is obtained by employing a norm that is generated by a complex valued "inner product $(\cdot\,,\cdot)$" defined for pairs of elements. The norm of an element x is given by:

$$||x|| = (x, \bar{x})^{1/2} \tag{3.25}$$

in which the inner product is assumed to satisfy the following conditions:

$$(x, \bar{x}) \geqslant 0 \quad \text{with } (x, \bar{x}) = 0 \text{ only for } x = 0 \tag{3.26}$$

and

$$(\alpha x + \beta y, z) = \alpha(x, z) + \beta(y, z) \qquad \text{(a)}$$
$$(x, y) = \overline{(y, x)} \qquad \text{(b)} \qquad (3.27)$$
$$|(x, y)| \leqslant ||x|| \cdot ||y|| \qquad \text{(c)}$$

for elements $x, y, z \in X$ and complex numbers α, β. The above conditions refer to a complex "Hilbert space". For a real Hilbert space the inner product is real valued, the scalars α, β are real and (3.27 b) is then replaced by $(x, y) = (y, x)$.

A "linear functional or operator M" defined on a Banach space X is a real (or complex) valued function defined for each element of X satisfying:

$$M(\alpha x + \beta y) = \alpha M(x) + \beta M(y) \tag{3.28}$$

for each pair of elements $x, y \in X$ and each pair of real (complex) numbers. The functional is bounded, if

$$|M| = \sup_{\substack{x \in X \\ x \neq 0}} \frac{|M(x)|}{||x||} < \infty \tag{3.29}$$

A bounded linear operator T on the Banach space X is called a "projection", if $T^2 = T$. The operator T determines also the subspaces X_1, X_2 of X with $X_1 = \{x \in X;\ Tx = x\}$, $X_2 = \{x \in X;\ Tx = 0\}$. T is called a "contraction operator", if $||T|| < 1$.

Of particular interest for the subsequent analysis concerned with the "deformation and stress space" of a structured solid, is the set of "locally convex" spaces. In this case every open set containing 0 contains a convex open set containing 0. Hence the topology of such vector spaces is given by a set of "semi-norms". A semi-norm p on a vector space X is a non-negative real function on X such that:

$$\begin{aligned} &p(x+y) \leqslant p(x)+p(y) \quad \text{for every } x, y \in X \\ &p(\alpha x) = |\alpha| p(x) \quad \text{for every } \alpha \in R \text{ and } x \in X \end{aligned} \tag{3.30}$$

Considering a convex open set U containing 0, then the set $V = U \cap (-U)$ is also a convex set containing 0. Hence, it is readily seen that for every $x \in X$ there is an $\alpha \in R^+$ such that $x \in \alpha V$, iff $-x \in \alpha V$. Furthermore, letting

$$\begin{aligned} &p(x) = \sup[\alpha:\ x \notin \alpha V, \alpha \geqslant 0], \quad \text{if } x \neq 0 \\ &p(0) = 0 \end{aligned} \tag{3.31}$$

it can be shown, that p is a semi-norm and that the sets

$$U_{p,r} = [x;\ p(x) < r] \tag{3.32}$$

for all p obtained in this manner and all $r > 0$ form a base for the topology in X at 0. Hence in a "locally convex topological vector space" the topology is given by a set p_α, $\alpha \in R^+$ of semi-norms.

Of further significance are the locally convex topological vector spaces that are "metrizable". Since a metrizable space is such that at each point of the space a base for the topology at that point is countable, it is apparent that the topology of a metrizable locally convex topological vector space can be given in terms of countably many semi-norms. Conversely, if X is such a space and where the topology is given by a countable number of semi-norms $\{p_n\}$ the space X is metrizable. Considering for instance semi-norms $\{p_n\}$ such that if $p_n(x) = 0$ for $n = 1, 2, \ldots$, then $x = 0$ and considering the metric of the form:

$$d(x, y) = \sum_{n=1}^{\infty} \frac{1}{2^n} \frac{p_n(x-y)}{1+p_n(x-y)} \tag{3.33}$$

it can be shown (see also reference[61]) that this metric is the same as the one given by the semi-norms $\{p_n\}$.

A wide class of spaces of a special type are known as "FK-spaces". An FK-space X is defined as follows:

(i) A vector space over the real (or complex) numbers whose elements are sequences of real (complex) numbers.

(ii) X is a locally convex metrizable topological vector space in which the topology is given by a countable set $\{p_n\}$ of semi-norms.

(iii) The metric space is complete.

A "complete metrizable locally convex topological vector space" is referred to as a "Fréchet space". Such spaces have many properties in common with Banach spaces and will be further discussed in subsequent sections of this chapter. For a more detailed study of such spaces the reader is referred to references[62–63].

3.3 General theory of stochastic deformations

The foregoing section dealt briefly with the structure of topological spaces and, in particular, that of topological vector spaces. In order to develop a general theory of stochastic deformations it is further necessary to consider some functional analytic aspects of this theory as given below.

(A) *Functional analytic aspects of deformation and stress*

In discussing the basic concepts or postulates of probabilistic micromechanics of structured solids (Chapter II, Section 2.3 B), it has been stated that for the analysis of the deformational behaviour of such media the existence of a set of admissible "state vectors" of the microelements at a given time is assumed. Hence the mechanical states of the individual elements of the microstructure are representable by an r-dimensional state vector ${}^{\alpha}\mathbf{v}$: ${}^{\alpha}v_i$ $(i = 1, 2, \ldots, r)$,

($\alpha = 1, 2, \ldots, N$); $\alpha \in Z^+$. Since the macroscopic material body is regarded as a collection of a finite number of non-intersecting mesodomains, two other sets of state vectors can be written (eqns. (2.7), (2.5), Chapter II). The set of all state vectors ${}^{\alpha}\boldsymbol{v}$ generates the "state space X".

However, due to experimental constraints, it is impossible to determine any state vector accurately and hence the latter can only be assessed within a certain range expressed by ${}^{\ni}\boldsymbol{v} < {}^{\alpha}\boldsymbol{v} < {}^{\ni}\boldsymbol{v} + \Delta{}^{\ni}\boldsymbol{v}$. The mechanical states of a microelement are therefore specified within this range by an accuracy $\Delta{}^{\ni}\boldsymbol{v}$ and ${}^{\ni}\boldsymbol{v}$ designates a specific value of ${}^{\alpha}\boldsymbol{v}$. If such a range is present for a given experiment the above statement means that one can at most specify only how many microelements α will be in their respective mechanical states within this range. Thus one obtains subsets of the state space X which can be regarded as open sets or spheres in the sense discussed in the preceding section (Axioms 3.10) and earlier in Chapter II (eqn. (2.6)). Denote by $\mathscr{F} = \{{}^{\ni}E\}$ the σ-algebra generated by the family of all multidimensional intervals in the state space X (open, semi-open, closed); in particular the intervals ${}^{\ni}E = \{{}^{\ni}\boldsymbol{v} < {}^{\alpha}\boldsymbol{v} < {}^{\ni}\boldsymbol{v} + \Delta{}^{\ni}\boldsymbol{v}\}$ and ${}^{\ni}E = \{{}^{\alpha}\boldsymbol{v} \leqslant {}^{\ni}\boldsymbol{v}\}$ belong to $\mathscr{F}$. The family $\mathscr{F}$ has the following properties:

$$\begin{aligned} &\text{(i) } {}^{\ni}E \in \mathscr{F} \Leftrightarrow {}^{\ni}\bar{E} \in \mathscr{F} \\ &\text{(ii) } {}^{\ni}E \in \mathscr{F} \Rightarrow \bigcup {}^{\ni}E \in \mathscr{F} \\ &\text{(iii) } X \subset \mathscr{F} \end{aligned} \tag{3.35}$$

The elements ${}^{\ni}E$ of the σ-algebra $\mathscr{F}$ are Borel sets of X and the space X becomes a measurable space $[X, \mathscr{F}]$ in the sense of Halmos[64].

In general, the state vector ${}^{\alpha}\boldsymbol{v}$ may have several component vectors such as the kinematic one ${}^{\alpha}\boldsymbol{\omega}$ (Chapter II), the stress vector ${}^{\alpha}\boldsymbol{\sigma}$ associated with the internal microstress and the surface stress within the interaction zone between microelements and others. Thus one can write ${}^{\alpha}\boldsymbol{v}$ as follows:

$${}^{\alpha}\boldsymbol{v} = \begin{bmatrix} {}^{\alpha}\boldsymbol{\omega} \\ {}^{\alpha}\boldsymbol{\sigma} \\ \vdots \end{bmatrix} \tag{3.36}$$

in which each of the component vectors belongs to a subspace of X. For instance, the vector ${}^{\alpha}\boldsymbol{\omega}$ belongs to the general kinematic subspace Ω (Chapter II) and includes amongst other quantities the microdeformations ${}^{\alpha}\boldsymbol{w}$ and ${}^{\alpha}\boldsymbol{\mu}$ or ${}^{\alpha\beta}\boldsymbol{d}$. Considering now that ${}^{\alpha}\boldsymbol{w} \in {}^{\alpha}U$ and ${}^{\alpha}\boldsymbol{\mu}$ or the relative distance vector ${}^{\alpha\beta}\boldsymbol{d} \in {}^{\alpha\beta}U$, then:

$$ {}^{\alpha}U \cup {}^{\alpha\beta}U = U \subset \Omega \tag{3.37} $$

where U is the space of all microdeformations ${}^{\alpha}\boldsymbol{u}$ (discussed in Chapter II). Since the analysis in this section is concerned with the functional analytic aspects of deformation and stress, then by considering the deformation space U first, the associated vector quantities will be denoted without a superscript. It is to be noted however, that such quantities always refer to the microelements ($\alpha = 1, 2, \ldots, N$) or the boundary zone between two elements α and β as the case may be. Denote by ${}^{\ni}E$ the interval $\{{}^{\ni}\boldsymbol{v} < {}^{\alpha}\boldsymbol{v} < {}^{\ni}\boldsymbol{v} + \Delta^{\ni}\boldsymbol{v}\}$. The following theorem may be stated:

T.1: The subspace of X is measurable, if and only if the set ${}^{\ni}E$ is the union of disjoint open spheres ${}^{\ni}E^{(k)}$ ($k = 1, 2, \ldots, n$). (3.38)

The proof of the above theorem and its corollary, e.g. that the deformation spaces ${}^{\alpha}U$, ${}^{\alpha\beta}U$ and U are all measurable spaces was given by Basu[65]. Thus, if the subsets of X of the form X_k ($k = 1, 2, \ldots, n$) have the elements $\boldsymbol{v}^{(k)} \subset \boldsymbol{v}$ and if the $\boldsymbol{v}^{(k)}$'s are such that the sets ${}^{\ni}E^{(k)}$ are disjoint, the σ-algebra $\mathscr{F}$ can be divided into subclasses $\mathscr{F}^{(k)}$ such that:

$$ \bigcup \mathscr{F}^{(k)} = \mathscr{F}, \quad \text{whenever} \bigcup_{k=1}^{n} {}^{\ni}E^{(k)} = \sum_{k=1}^{n} {}^{\ni}E^{(k)} = {}^{\ni}E \tag{3.39} $$

Whilst this theorem shows the spaces to be measurable a proper probability measure must be introduced as well. The distribution function of a random variable defines uniquely a probability measure on the space of values of this variable. For instance, the distribution function $\mathscr{P}({}^{\ni}\boldsymbol{v})$ corresponding to state ${}^{\alpha}\boldsymbol{v}$ defines the measure $\mathscr{P}\{{}^{\ni}I\}$ on the state space X as follows:

$$ \mathscr{P}\{{}^{\ni}I\} = \mathscr{P}({}^{\ni}\boldsymbol{v}) \qquad \text{for } {}^{\ni}I = \{{}^{\alpha}\boldsymbol{v}: {}^{\alpha}\boldsymbol{v} \leqslant {}^{\ni}\boldsymbol{v}\}. $$

Taking this function as a measure on $[X, \mathscr{F}]$ one obtains the probability space or equivalently the measure space $[X, \mathscr{F}, \mathscr{P}]$. Alternatively, one may consider the sets ${}^{\ni}E$ as events of a random experiment carried out to determine the mechanical states of an ensemble of microelements at a specific time. Then ${}^{\ni}E$ would represent the event that the mechanical state of the element α is within $({}^{\ni}\boldsymbol{v}, {}^{\ni}\boldsymbol{v}+\Delta^{\ni}\boldsymbol{v})$ and from Theorem 1 one could consider the range of microdeformations ${}^{\alpha}\boldsymbol{u}$ only, so that:

$$ {}^{\ni}E(\boldsymbol{u}) = \mathscr{P}\{{}^{\ni}\boldsymbol{u} < {}^{\alpha}\boldsymbol{u} < {}^{\ni}\boldsymbol{u}+\Delta^{\ni}\boldsymbol{u}\} \tag{3.40} $$

The corresponding probability measure is given by:

$$ \begin{aligned} &\mathscr{P}^{u}\{{}^{\ni}I(\boldsymbol{u})\} = \mathscr{P}^{u}({}^{\ni}\boldsymbol{u}), \qquad {}^{\ni}I(\boldsymbol{u}) = \{{}^{\alpha}\boldsymbol{u}: {}^{\alpha}\boldsymbol{u} \leqslant {}^{\ni}\boldsymbol{u}\}, \qquad \ni \in Z^{+} \\ &\mathscr{P}^{u}(U) = 1 \end{aligned} \tag{3.41} $$

It should be noted that due to the accesibility of distribution functions in experimental micromechanics, it is more convenient to consider in the theory of stochastic deformations a "cumulative measure", which can be defined in terms of the distribution function of the relevant quantities.

In accordance with earlier statements the subspace U with the σ-algebra $\mathscr{F}^{u}\{{}^{\ni}E(\boldsymbol{u})\}$ generated by the intervals of the space U (similarly to σ-algebra $\mathscr{F}$ considered before) and $\mathscr{P}^{u}$ defined above, constitute a probability space referred to in this work as the "deformation space" $[U, \mathscr{F}^{u}, \mathscr{P}^{u}]$. The measure $\mathscr{P}^{u}$ will be called the "cumulative measure". If $p^{u}(\boldsymbol{u})$ is the density of $\mathscr{P}^{u}(\boldsymbol{u})$, then the measure $\mathscr{P}^{u}\{E\}$ is representable in the form

$$ \mathscr{P}^{u}\{E\} = \sum_{u \in E} p^{u}(\boldsymbol{u}) \qquad \text{for } E \in \mathscr{F}^{u} $$

or (3.42)

$$ \mathscr{P}^{u}\{E\} = \int_{E} \mathrm{d}\mathscr{P}^{u}(\boldsymbol{u}) = \int_{E} p^{u}(\boldsymbol{u})\,\mathrm{d}\boldsymbol{u} \qquad \text{for } E \in \mathscr{F}^{u} $$

for the discrete and continuous cases, respectively.

It is evident that, in an analogous manner, subspaces associated with the deformations ${}^{\alpha}\boldsymbol{w}$ and ${}^{\alpha}\boldsymbol{u}$ or ${}^{\alpha\beta}\boldsymbol{d}$ can be formed, represented by $[{}^{\alpha}U, \mathscr{F}^{\alpha w}, \mathscr{P}^{\alpha w}]$ and $[{}^{\alpha\beta}U, \mathscr{F}^{\alpha\beta d}, \mathscr{P}^{\alpha\beta d}]$, respectively. From a knowledge of these spaces the more general space U can then be constructed. Considerations of the latter space in the formulation of a general deformation theory is highly significant, since present experimental methods permit the determination of the microdeformations ${}^{\alpha}\boldsymbol{u}$ (see also Chapter VI). However, new techniques are at present under investigation to obtain information on the relative distance vector ${}^{\alpha\beta}\boldsymbol{d}$.

A second theorem has been stated and proved in reference[65] concerning the regularity of the above probability measure $\mathscr{P}^u$. The result is of great significance in the present theory since one can then define a random vector $\boldsymbol{u} \in U$ as a regular measurable function in the space U. Whilst the concept of the above function space is mathematically convenient in the theory of stochastic deformation of structured media, it is further necessary to introduce a set of norms or semi-norms in the corresponding probability space that comply with the physical problem under consideration.

Thus, if U is the space of all $\mathscr{P}^u$-regular measurable and bounded functions of $\boldsymbol{u}$, which may be discrete or continuous, then the expected value of $\boldsymbol{u}$ in these cases becomes:

$$E\{\boldsymbol{u}\} = \sum {}^{\ni}\boldsymbol{u}p({}^{\ni}\boldsymbol{u}) = \sum {}^{\ni}\boldsymbol{u}\Delta\mathscr{P}^u({}^{\ni}\boldsymbol{u})$$

and (3.43)

$$E\{\boldsymbol{u}\} = \int {}^{\ni}\boldsymbol{u}\,\mathrm{d}\mathscr{P}^u({}^{\ni}\boldsymbol{u})$$

where $E\{\boldsymbol{u}\} = \langle\boldsymbol{u}\rangle$ is the mean value of the random deformation vector. It is to be noted, that if in an experiment the mean value $\langle\boldsymbol{u}\rangle$ which is representative of a macroscopic deformation is equal to zero, although microdeformations within the structured medium are not zero valued, the absolute value of $|\langle\boldsymbol{u}\rangle|$ or of $|E\{\boldsymbol{u}\}|$ will only satisfy the properties of semi-norms (see also Yoshida[63]). Hence the topological vector space U will be a Fréchet space with the semi-norm defined in the above manner. On the other hand, if in an experiment one is only concerned with determining the average value of the magnitude of the microdeformations $\boldsymbol{u}$, an analogous definition to that given in (3.43) would satisfy the properties required of a norm and hence the space would then become a Banach space.

One can give other definitions of semi-norms. For example, using the "standard deviation of the microdeformation" and again considering U to be the space of all $\mathscr{P}^u$-regular measurable and bounded functions $\boldsymbol{u}$, then for the discrete and continuous case, respectively:

$$D^u = \left\{\sum |^{\ni}\boldsymbol{u} - \langle \boldsymbol{u} \rangle|^2 \Delta \mathscr{P}^u\right\}^{1/2} \qquad \text{(a)}$$

and (3.44)

$$D^u = \left\{\int |^{\ni}\boldsymbol{u} - \langle \boldsymbol{u} \rangle|^2 \mathrm{d}\mathscr{P}^u\right\}^{1/2} \qquad \text{(b)}$$

from which it may be seen that $|D\{\boldsymbol{u}\}|$ satisfies the properties of a semi-norm and that the U space will be a Fréchet space (see also Bourbaki[66], Treves[62], Yoshida[63] and Kappos[19]).

Finally, one can introduce other definitions of the probability measure in accordance with the definition of probability. Thus a more fundamental measure is that of the "conditional probability measure" as discussed in the following section.

In order to complete this section concerned with the functional analytic aspect of deformation and stress, the stress vector $^{\alpha}\boldsymbol{\sigma}$ as a component vector of $^{\alpha}\boldsymbol{v}$ and the corresponding "stress space Ξ" it generates will be considered below. It has been mentioned earlier that $^{\alpha}\boldsymbol{\sigma}$ itself consists of two components, the internal stress $^{\alpha}\xi$ and the surface stress in the interaction zone between elements $^{\alpha\beta}\xi$, thus:

$$^{\alpha}\boldsymbol{\sigma} = \begin{bmatrix} ^{\alpha}\xi \\ ^{\alpha\beta}\xi \end{bmatrix} \tag{3.45}$$

Considering now that $^{\alpha}\boldsymbol{\sigma}$ belongs to the subspace Ξ of X, then each of the random stresses $^{\alpha}\xi$ and $^{\alpha\beta}\xi$ belong to corresponding subspaces of Ξ, viz:

$$^{\alpha}\xi \in {^{\alpha}\Xi}, \quad ^{\alpha\beta}\xi \in {^{\alpha\beta}\Xi}, \quad ^{\alpha}\Xi \cup {^{\alpha\beta}\Xi} = \Xi \subset X \tag{3.46}$$

Following the arguments given earlier that all subspaces and Ξ are measurable, then a σ-algebra may be defined in these spaces completely analogous to that of the deformation space. Hence, in this case the following set is a Borel set:

$$^{\ni}E = \{^{\ni}\xi < {^{\alpha}\xi} < {^{\ni}\xi} + \Delta^{\ni}\xi\}, \quad \ni \in Z^+ \tag{3.47}$$

where $^{\ni}\xi$ designates a particular value of the random microstress $^{\alpha}\xi$ and $\Delta^{\ni}\xi$ the accuracy with which such a stress could be assessed. It should be noted that it is possible to use a more general tensor quantity $^{\alpha}\boldsymbol{\sigma}$, which includes both the internal microstress $^{\alpha}\xi$ within a microelement with reference to the attached body frame of that element and the surface stress $^{\alpha\beta}\xi$ taken with reference to the surface coordinate frame. Hence the components of the two stress tensors include the orientation of these reference frames with respect to the external coordinate systems. This will be more fully discussed in the following chapter, when dealing with the response behaviour of polycrystalline solids. Thus the random stress tensor ${}^{\alpha}\boldsymbol{\sigma}$ referred to subsequently as an element of the space Ξ can be obtained from the stress vector ${}^{\alpha}\boldsymbol{\sigma}$ in the above described manner.

The σ-algebra $\mathscr{F}^{\sigma}$ in Ξ contains for instance all sets of the form

$$ {}^{\ni}E = \{{}^{\ni}\boldsymbol{\sigma} < {}^{\alpha}\boldsymbol{\sigma} < {}^{\ni}\boldsymbol{\sigma} + \Delta^{\ni}\boldsymbol{\sigma}\}, \qquad \ni \in Z^{+} \tag{3.48} $$

To construct the probabilistic structure of the stress space Ξ a cumulative probability measure on $\mathscr{F}^{\sigma}$ can be written as:

$$ \begin{aligned} &\mathscr{P}^{\sigma}\{{}^{\ni}E\} = \mathscr{P}^{\sigma}({}^{\ni}\boldsymbol{\sigma}) \\ &{}^{\ni}E = \{{}^{\alpha}\boldsymbol{\sigma} \leqslant {}^{\ni}\boldsymbol{\sigma}\}, \qquad \ni \in Z^{+} \end{aligned} \tag{3.49} $$

However, it is important to note that the choice of this probability measure is not arbitrary, since by choosing *a priori* the probability measure on the deformation space U, the choice of the measure on the stress space Ξ is dictated by the probability distribution of the "material operator". This will be further clarified in Section 3.5 of this chapter. Thus under the above restriction one can construct as before a probabilistic function space pertaining to the microstress denoted by the triplet $[\Xi, \mathscr{F}^{\sigma}, \mathscr{P}^{\sigma}]$.

It is of interest to note that the state vector ${}^{\alpha}\boldsymbol{v}$ can be regarded for the simplification of the analysis to consist of the components ${}^{\alpha}\boldsymbol{u}$ and ${}^{\alpha}\boldsymbol{\sigma}$ only, since in both these quantities other thermo-mechanical parameters may be included. One could further consider the microstrains ${}^{\alpha}\boldsymbol{\varepsilon}$ and ${}^{\alpha\beta}\boldsymbol{\varepsilon}$ in the random variable ${}^{\alpha}\boldsymbol{\omega}$ as gradient functions of ${}^{\alpha}\boldsymbol{w}$ and ${}^{\alpha\beta}\boldsymbol{d}$ so that the more general state space X may be thought of being composed of the spaces U and Ξ only. The visualization of this representation of the behaviour of the physical system permits the latter to be characterized by an input-output relationship. Thus, for example, the input may be regarded as an element of Ξ and the output as an element of U. It can be shown that for a physical system a mapping between Ξ and U always exists and that the topological structure of one of the spaces, for instance that of Ξ, may be given in terms of a non-degenerate bilinear form with respect to this mapping (see also

Moreau[67] and Tonti[68]). This however will involve the introduction of a material operator as mentioned previously. Alternatively, one can consider the spaces Ξ and U independently, e.g. without reference to the existence of a material operator that links these two spaces. The topological structure of Ξ will then be induced by the nature of the physical problem under consideration. Thus, if a set of elements ${}^{\alpha}\sigma \in \Xi$ with a locally convex topology defined by the open sets as given by (3.48) is considered and taking the related measure in the form of (3.49), the following semi-norm may be defined:

$$|E\{{}^{\alpha}\sigma\}| = \left|\sum {}^{\ni}\sigma \Delta \mathscr{P}^{\sigma}({}^{\ni}\sigma)\right| \tag{3.50}$$

in the case of a discrete random variable ${}^{\alpha}\sigma$ and if the latter is continuous, the semi-norm will be:

$$|E\{{}^{\alpha}\sigma\}| = \left|\int {}^{\ni}\sigma \, \mathrm{d}\mathscr{P}^{\sigma}({}^{\ni}\sigma)\right| \tag{3.51}$$

so that the associated space $[\Xi, \mathscr{F}^{\sigma}, \mathscr{P}^{\sigma}]$ has the structure of a Fréchet space. It is equally possible to use the standard deviation of ${}^{\alpha}\sigma$ in both the discrete and continuous case and on the assumption of a uniform stress distribution one obtains the same topology, i.e. the representation of the stress space by the above triplet.

However, of more general interest is the case of a non-uniform stress distribution, which is more likely to occur in most practical applications and for which $|D^{\sigma}|$ will not be zero, although some of the elements of Ξ may be zero valued. For such a case, it is not possible to define a norm or semi-norm in the above manner and a more suitable definition for the probability measure must be established. Thus depending on the definition for an appropriate measure, different topologies will arise for the representation of the stress space Ξ. However, for the purpose of developing the more general theory of stochastic deformations as discussed below, it is convenient to restrict the considerations to the Fréchet space structure of the stress space.

(B) *General theory of stochastic deformations*

In the foregoing paragraphs of this section, the functional analytic aspects of deformation and stress and the topological structure of the associated function spaces have been briefly discussed. In order to develop a general theory of stochastic deformations, which is one of the main objectives of the present work, it will be necessary to clarify the concept of a deformation and stress field pertaining to the physical domain of a structured solid.

Thus considering a set of function spaces $[U, \mathscr{F}^u, \mathscr{P}^u]$ representing the general deformation space, then each of these spaces corresponds to a point on the real line R^+. In particular, if $R^+ = [0, \infty)$ in which an element $t \in R^+$, the occurring deformations at this instant of time t can be regarded as a random variable in the space $[U, \mathscr{F}^u, \mathscr{P}^u]$.

However, in a more general sense the deformation $\boldsymbol{u}(X; t)$ designates a "random deformation process", where the parameter t refers to time and belongs to the positive half of the real line, R^+. Hence for a fixed time $t \in R^+$, the deformation $\boldsymbol{u}(X)$ is a random deformation vector in the space $[U, \mathscr{F}^u, \mathscr{P}^u]$. It is to be noted that the function $\boldsymbol{u}(X; t)$ is such that for a fixed time t, $\boldsymbol{u}(X)$ is a random variable or, in general, referred to as a random function. The theory of random processes is thus related to the theory of random functions in the appropriate function spaces. Following Halmos[64], it is possible to consider an infinite number of such spaces and to construct an infinite product space $[U^\infty, \mathscr{F}^\infty, \mathscr{P}^u]$ where the superscript "∞" indicates that an n-fold product space can be extended theoretically to ∞, if the function $\boldsymbol{u}(X; t)$ is considered as a t-continuous random function. Alternatively one may consider the function space $[U, \mathscr{F}^u, \mathscr{P}^u]$ and a one-parameter family of mappings T_t such that:

$$T_t\colon U \to U \text{ for all } t \in R^+ \qquad \boldsymbol{u}_t(X) \in U \tag{3.52}$$

so that the random process of deformation is characterized by a measurable function $\boldsymbol{u}(X; t)$ or $\boldsymbol{u}_t\,(X)$.

It is to be noted that the σ-algebra $\mathscr{F}^\infty$ of this product space has a countably finite number of Borel sets corresponding to the sequence $t_1, t_2, \ldots, t_n \in R^+$, whereby each set is obtainable from the other by the automorphism T_t or its inverse. In this context the theorem stated in the work of Basu[65] is important. This theorem is as follows:

T.1: If $\mathscr{P}^u$ is a regular measure in the open set E_r, for instance, and T_t a mapping defined by (3.52), then $T_t E_r(\boldsymbol{u}) = E_{r+1}(\boldsymbol{u})$, $r = 1, 2, \ldots, n$ and $E_{r+1}(\boldsymbol{u})$ is $\mathscr{P}^u$-regular measurable.

The theorem implies that if $\mathscr{P}^u$ is a regular measure on the set $E_r(\boldsymbol{u})$, then $E_{r+1}(\boldsymbol{u})$ is also $\mathscr{P}^u$-regular measurable. The result of the above theorem can be generalized for any Borel sets $E_r(\boldsymbol{u})$; $r = 1, 2, \ldots, n$ in $\mathscr{F}^\infty$ and hence it may be concluded that at any instant of time during a general deformation process, the random deformation is $\mathscr{P}^u$-measurable. Hence, when considering a specific deformation phenomenon as dealt with later within the framework of the general theory of stochastic deformations, it is fundamental to specify the appropriate random process and the associated probability measure. Thus in a deformation process describing the purely elastic response of a structured medium, the probability measure is time-independent. The corresponding random process is then strictly stationary, if the probability measures on all Borel sets $E_r \in \mathscr{F}^u$ are equal, e.g.:

$$\mathscr{P}^u\{E_{r+1}\} = \mathscr{P}^u\{E_r\}, \qquad r = 1, 2, \ldots, n \tag{3.53}$$

where

$$E_{r+1} = T_t E_r$$

Random processes of this type are discussed amongst others by Bochner[69] and Doob[9].

Whilst the time-independent probability measure allows consideration of the purely elastic deformation behaviour of a structured solid, the time-dependent measure applicable to the inelastic response is more significant. Considerations to this effect require, however, the introduction of a "conditional probability measure" corresponding to the concept of conditional probability outlined in Chapter I of this text. This concept is mainly due to Kolmogorov[2] and Rényi[5]. This measure may be defined on the basis of the properties of an extended real-valued set function in $\mathscr{F}^\infty$, which are as follows:

$$\begin{aligned}
&\text{(i) } 0 \leqslant \mathscr{P}\{E_i|E_j\} \leqslant 1, \quad \forall E_i, E_j \in \mathscr{F}^\infty, \quad i \neq j \\
&\text{(ii) } \mathscr{P}\{E_i|E_j\} = 0 \Leftrightarrow E_i \cap E_j = \emptyset, \quad i \neq j \\
&\text{(iii) } \mathscr{P}\{E_i|E_i\} = 1 \\
&\text{(iv) } \mathscr{P}\{U|E_i\} = 1
\end{aligned} \tag{3.54}$$

The conditional probability measure $\mathscr{P}\{E_i|E_j\}$ satisfying the above properties (3.54) defines now a probability measure $\mathscr{P}\{E_i(\boldsymbol{u})\}$ with the condition that the event corresponding to the Borel set E_j is certain. This statement is of considerable importance since, if in an experiment the initial probability distribution of the microdeformations can be established, one can then construct a deformation process from the knowledge of $\mathscr{P}\{E_i|E_j\}$, analytically. Within the framework of a general theory of stochastic deformations this conditional probability measure is of fundamental importance, since the following theorem can be stated:

T.2: To each automorphism T_t there corresponds a conditional probability measure $\mathscr{P}\{E_{r+1}|E_r\}$ such that, whenever T_t: $E_r = E_{r+1}$

$$\mathscr{P}^u\{E_{r+1}\} = \mathscr{P}^u\{E_{r+1}|E_r\}\mathscr{P}^u\{E_r\} \tag{3.55}$$

The proof of the above theorem is given in reference[65] in which a generalization of it to a set of conditional probability measures is also considered. Thus $\mathscr{P}\{E_{r+1}|E_r\}$, $r = 1, 2, \ldots, n-1$ is such that:

$$\mathscr{P}^u\{E_n\} = \mathscr{P}^u\{E_1\}\prod_{r=1}^{n-1}\mathscr{P}\{E_{r+1}|E_r\} \tag{3.56}$$

Rewriting relation (3.55) in the following form:

$$\mathscr{P}^u\{E_{r+1}, t_{r+1}\} = \mathscr{P}\{t_{r+1}, t_r\}\mathscr{P}^u\{E_r, t_r\} \tag{3.57}$$

where E_r corresponds to $t_r \in R^+$ and E_{r+1} to $t_{r+1} > t_r$, it is readily seen that by considering a deformation process for which $\mathscr{P}\{t_{r+1}, t_r\}$ depends only on the time difference, the sequence of deformations $\{\boldsymbol{u}_t(X)\}$ belonging to the space $[U, \mathscr{F}^u, \mathscr{P}^u]$ together with the conditional probability measure $\mathscr{P}\{t_{r+1}, t_r\}$ describes indeed a homogeneous Markov process. Furthermore, denoting t_{r+1} simply by t and t_r by 0, equation (3.57) can be written as:

$$\mathscr{P}^u\{t\} = \mathscr{P}\{\boldsymbol{u}(t)\} = \mathscr{P}(t)\mathscr{P}^u\{0\} = \mathscr{P}(t)\mathscr{P}\{\boldsymbol{u}(0)\} \tag{3.58}$$

indicating the possibility of computing the distributions of microdeformations at any time t, if the initial distribution $\mathscr{P}\{\boldsymbol{u}(0)\}$ can be established. Experimental methods dealing with this aspect of the deformation theory will be discussed in the last chapter of this text. It will be seen subsequently that an analogous form to (3.58) can be obtained for the distribution of microstresses. The latter arises from the same assumption, e.g. that the deformation process can be regarded as a special class of random processes. In particular, for the purely elastic and steady-state deformations a structured solid may undergo, the deformation process is approximated by a Markov process. The intermediary or transient state between these two states is representable by a quasi-Poisson process as shown later. For the development of a general deformation theory these three stages of deformation must be taken into account in a unified manner.

Continuing the above considerations regarding the general deformation theory and considering the special class of random processes, e.g. the Markov processes, then the space of all measures $\mathscr{P}^u\{E_r\}$, $r = 1, 2, \ldots$ can be denoted by $L(0, 1)$. The conditional probability defined by (3.55) can be regarded as an operator on $L(0, 1)$ to $L(0, 1)$. It is however conventionally referred to as transition probability. Following Dynkin[29] and Bharucha-Reid[30] for instance and considering a closed interval of time $[t, s] \in R^+$, subdividing it into smaller intervals and selecting a point $\tau > t$; $\tau \in [t, s]$ shows that the transition probability satisfies the Chapman–Kolmogorov functional relation, i.e.:

$$\boldsymbol{P}\{t, s\} = \int_U \boldsymbol{P}\{t, \tau\} \mathrm{d}\boldsymbol{P}\{\tau, s\} \tag{3.59}$$

The significance of the above transition probability and its corresponding matrix in the general deformation theory will be discussed later. Relation (3.59) connects the stochastic theory with the functional analytic formulation as pointed out earlier.

It should be noted however, that the above choice of random processes, e.g. the Markov processes to represent the deformational behaviour of structured media leads to a restriction of the class of materials that can be treated by the formulation, since the interval between two successive distributions as obtained from experiments, must be small. This restriction leads to consideration of the steady-state deformations in terms of a homogeneous Markov type process for which relation (3.59) is given by:

$$\boldsymbol{P}\{t-s\} = \int_U \boldsymbol{P}\{t-\tau\} \mathrm{d}\boldsymbol{P}\{\tau-s\} \tag{3.60}$$

Relation (3.59) can also be written in matrix form:

$$\boldsymbol{P}(t+s) = \boldsymbol{P}(t)\boldsymbol{P}(s) \tag{3.61}$$

indicating clearly that the system has a semi-group property. In this context it is of interest to note, that classical statistical mechanics relates this semi-group property to the notion of irreversibility of the system (see for instance Kampé de Fériet[70] and Birkhoff, Bona and Kampé de Fériet[53]).

Considering now a "time homogeneous" process, which depends on the difference of time $(t-s)$ only and following the discussion given in reference[9], the transition matrix representing a transition from state "i" to state "j" denoted by $\boldsymbol{P}_{ij}(t)$ is obtained as follows[71]:

$$\lim_{t\to 0^+} \frac{1-\boldsymbol{P}_{ii}(t)}{t} = -q_{ii} < \infty \tag{3.62}$$

and

$$\lim_{t\to 0^+} \frac{\boldsymbol{P}_{ij}(t)}{t} = q_{ij}(t) \tag{3.63}$$

for a σ-measurable function and for all $t \in R^+$.

It should be noted that a time-homogeneous process is equivalent to a process in which the stress boundary conditions of the problem are time-independent.

It has been shown in reference[65] that the transition probability in the case of a steady-state deformation leads to properties, which permits such a process to be defined in the following manner:

(i) $\boldsymbol{P}(t)$ is a contraction on $L(0, 1)$ to $L(0, 1)$ such that $||\boldsymbol{P}(t)|| \leqslant 1$.

(ii) $\forall\, s, t \in R^+$ and $E \in \mathscr{F}^u$ of U corresponding to t

$$\lim_{t\to s} [\boldsymbol{P}(t)-\boldsymbol{P}(s)]\mathscr{P}\{E\} = 0 \tag{3.64}$$

(iii) $\forall\, s, t \in R^+$ $\boldsymbol{P}(t+s) = \boldsymbol{P}(t)\boldsymbol{P}(s)$ with $\boldsymbol{P}(0) = I$.

Hence the Chapman–Kolmogorov functional equation becomes in terms of a matrix differential equation:

$$\frac{\mathrm{d}\boldsymbol{P}(t)}{\mathrm{d}t} = \boldsymbol{Q}(t)\boldsymbol{P}(t) \tag{3.65}$$

with

$$\boldsymbol{P}(0) = I, \quad \text{(identity matrix)} \tag{3.66}$$

It follows from the above relations ((3.65) and (3.66)) and the properties of the transition probability $\boldsymbol{P}(t)$ that in the case of elastic deformations only, the deformation process is time-independent and the corresponding random process a stationary one. In this case the stochastic deformation $\boldsymbol{u}(t)$ can be considered as belonging to the product space of countably finite copies of the function spaces $[U, \mathscr{F}^u, \mathscr{P}^u]$ with the corresponding measures on each of the spaces generated by the transition probabilities given by (3.55) and where the following properties of the transition probability will hold:

(i) The transition probability $\boldsymbol{P}(t, t+\Delta t)$ of the process $\boldsymbol{u}(t)$ changing from a state "i" to another state "j" in the interval Δt is zero. (3.67)

(ii) The transition probability $\boldsymbol{P}(t, t+\Delta t)$ of no change in the interval Δt is one.

For clarity, it should be noted that the symbols $\boldsymbol{u}(X, t)$ or $\boldsymbol{u}_t(X)$ introduced in the beginning of this section to represent a random microdeformation process has been replaced by $\boldsymbol{u}(t)$. Hence, it can be seen that the transition probability is determined by a constant matrix, i.e.:

$$\begin{aligned} \boldsymbol{P} &= \text{constant matrix} \\ \boldsymbol{Q} &= 0 \end{aligned} \tag{3.68}$$

so that the Kolmogorov differential equation reduces to:

$$\frac{\mathrm{d}\boldsymbol{P}}{\mathrm{d}t} = 0 \tag{3.69}$$

which can be solved for an initial value $\boldsymbol{P}(0) = I$ leading to the result:

$$\boldsymbol{P}(t) = I, \quad 0 < t \leqslant t_1 \tag{3.70}$$

The above relation suggests that the probability distribution of the stochastic microdeformations within the reversible or elastic response of the material remains constant with respect to time. In general, however, the transition probability $\boldsymbol{P}(t)$ and the transition matrix $\boldsymbol{Q}(t)$ which are both matrices whose elements are some probability measures, are arbitrary functions of time so long as relations (3.64) are satisfied. It is also of interest to note that the above result is consistent with that which is discussed below concerning the steady-state and the transient deformation process.

In discussing first the steady-state deformation process, the evolution of the distribution $\mathscr{P}\{\boldsymbol{u}(t)\}$ is again representable by the transition probability $\boldsymbol{P}(t)$ and the corresponding matrix $\boldsymbol{Q}(t)$. Thus, considering the random microdeformation vector $\boldsymbol{u}(t)$ changing from one state to another during the deformation, it can conveniently be characterized by a certain "intensity factor", which is directly related to the transition matrix and where it is assumed that the elements of this matrix remain time-independent within the time interval of two successive observations of deformation. For instance, if the deformation within a time interval Δt from an initial state "0" at $(t = 0)$ changes to a state "3" at $(t = \Delta t)$, then the intensity of this change is exactly the element q_{03} of the matrix $\boldsymbol{Q}$. Assuming now that Δt is small enough so that a change of $\boldsymbol{u}(t)$ from a state "i" to a neighbouring state "j" occurs only if $j = i+1$, $i = 0, 1, 2, \ldots$ then the following criteria for a steady-state deformation process may be given:

(i) The intensity with which the random deformation $\boldsymbol{u}(t)$ changes from a state "i" to its adjacent state "j" in the time interval Δt is $\lambda\Delta t + O(\Delta t)$, where λ is a positive constant and $O(\Delta t)$ the order of magnitude of Δt.

(ii) The intensity with which $\boldsymbol{u}(t)$ changes from a state "i" to any other state except "j", $j = i+1$ in the interval Δt is $O(\Delta t)$.

(iii) The intensity with which the random deformation $\boldsymbol{u}(t)$ does not change from a state "i" to a state "j" in Δt is $\{1-\lambda\Delta t - O(\Delta t)\}$.

It follows from these criteria that in the case of a steady-state deformation process, the elements of the transition matrix and the intensity are related as follows:

$$q_{ij} = \begin{cases} -\lambda & \text{for } i = 0, 1, \ldots \\ \lambda & \text{for } j = i+1 \\ 0 & \text{otherwise} \end{cases} \tag{3.71}$$

Hence the transition matrix $\boldsymbol{Q}$ can be written as:

$$\boldsymbol{Q} = \begin{bmatrix} -\lambda & \lambda & 0 & 0 & \ldots & 0 & 0 \\ 0 & -\lambda & \lambda & 0 & \ldots & 0 & 0 \\ \cdot & \cdot & \cdot & \cdot & \cdot & \cdot & \cdot \\ 0 & 0 & 0 & & \ldots & 0 & 0 \end{bmatrix} \tag{3.72}$$

It is evident from the above form of $\boldsymbol{Q}$ that the steady-state deformation process can be represented by a Poisson process. A more detailed study of such processes can be found in the work of Rényi[72], Prekopa[73] and Urbanik[74]. The probability densities are also discussed in depth by Bharucha-Reid[30]. It is of interest to note that the application of the notion of an intensity factor λ in the present theory is of utmost significance, since it indicates the "degree of transition" from one state to another. By this is meant, that if by a change of state certain microelements are undergoing stochastic deformations, then λ can be regarded as a measure of how many microelements actually take part in such a deformation within a specific mesodomain of the material body. Thus in contrast to the continuum mechanics point of view that implies that during the irreversible behaviour only positive deformations can occur, i.e. that the microdeformation vector $\boldsymbol{u}(t)$ cannot belong to the same state at two different times, the present theory still admits the probability, which is proportional to $(1-\lambda\Delta t)$, that the element remains in the same state.

Hence, using the above criteria, the Chapman–Kolmogorov relation for the steady-state deformation can be expressed by:

$$\frac{d\boldsymbol{P}_{ij}(t)}{dt} = -\lambda \boldsymbol{P}_{ij}(t) + \lambda \boldsymbol{P}_{i,j-1}(t), \qquad t_2 \leqslant t < \infty \tag{3.73}$$

in which t_2 refers to the instant of time at which the steady-state process is initiated. It has been shown in reference[30], that a solution of this matrix differential equation exists and is unique for a specified initial condition. Thus taking $\boldsymbol{P}(0) = I$ as an initial condition leads by employing an iteration procedure[30] to the following solution:

$$\boldsymbol{P}_{ij}(t) = \begin{cases} \dfrac{[\lambda(t-t_2)]^{j-i}}{(j-i)!} e^{-\lambda(t-t_2)} & \text{for } j \geqslant i, \quad t_2 \leqslant t < \infty \\ 0 & \text{for } j < i \end{cases} \tag{3.74}$$

So far as the time-dependent transition probabilities are concerned, which have been investigated in detail in reference[65], it is possible by changing the time variable from the above form to that of a closed interval $[t, s] \in R^+$ (see also eqn. (3.59)) to obtain the following result:

$$\sum \boldsymbol{P}_{ik}(t)\boldsymbol{P}_{kj}(s) = \frac{[\lambda(t+s)]^{j-i}}{(j-i)!} e^{-\lambda(t+s)} = \boldsymbol{P}_{ij}(t+s) \tag{3.75}$$

where $\boldsymbol{P}_{ik}$ and $\boldsymbol{P}_{kj}$ correspond to the transition probabilities of equation (3.61). It is readily seen from (3.74) that for $j = i+1$, one obtains the transition probability in the form of:

$$\boldsymbol{P}_{i,i+1}(t) = e^{-\lambda(t-t_2)}, \qquad t_2 \leqslant t < \infty \tag{3.76}$$

in which $\boldsymbol{P}_{i,i+1}(t)$ is frequently referred to as the "one-step transition probability" (see also Rosenblatt[75], Prohorov and Rozanov[6]). The "∞" sign on the right-hand of relation (3.73), (3.74) and (3.76) conforms to an "idealized" steady-state deformation process, which in actual systems will cease at a finite time that is related to the breakdown of the microstructure and the onset of fracture. In terms of the "one-step" transition probability, it is evident that the latter can be expressed in relation to two successive observations of microdeformations as follows:

$$\mathscr{P}^u\{t\} = e^{-\lambda(t-t_2)}\mathscr{P}^u\{t_2\} \tag{3.77}$$

where $\mathscr{P}^u\{t\}$ is obtainable from $\mathscr{P}^u\{t_2\}$ by means of relation (3.58). In general, the microdeformation distribution $\mathscr{P}^u\{t_2\}$ is multivariate and can be expressed in terms of the components of the vector valued random variable $\boldsymbol{u}(t)$. In this context as discussed later, it is more convenient to use marginal distribution functions or distribution functions of the individual components of $\boldsymbol{u}(t)$. The method of obtaining marginal distribution functions from the joint one, is well known and can be found in various texts given in the references of Chapter I (see for instance Pugachev[7]).

For completion of the formulation of a general theory of stochastic deformations with the exclusion of that part of the deformation process that is associated with the breakdown of a given microstructure, it remains to analyse the transient state of the material behaviour between the purely reversible and irreversible steady-state deformation.

Thus, considering the transition matrix which is essential in the formulation of the corresponding Kolmogorov differential equation, it is necessary for the transient phenomenon of deformations, that this matrix satisfies the following condition:

$$\begin{aligned} &\lim_{t\to t_1} \boldsymbol{Q}(t) \to \text{reversibility (elastic deformation)} \\ &\lim_{t\to t_2} \boldsymbol{Q}(t) \to \text{irreversibility (steady-state deformation)} \end{aligned} \tag{3.78}$$

in which the time instant t_1 is associated with the cessation of the purely elastic response and t_2 corresponds to the onset of a steady-state deformation, respectively. These time instances, if measurable, refer to the microstructural conditions of the material. Assuming thus that the two limiting conditions correspond to a fixed time interval $[t_1, t_2]$ of the general deformation process and subdividing it into incremental time intervals Δt in which t is a variable within $t_1 \leqslant t \leqslant t_2$ the probability of changing from a state "i" to an adjacent state "j" in this range can be stated as follows:

(i) The probability of a change in the interval $(t, t+\Delta t)$ is:

$$\left\{\frac{-\alpha t_1(t-t_2)+\beta t_2(t-t_1)}{t(t_2-t_1)}\right\}\Delta t+O(\Delta t), \qquad t_1 \leqslant t \leqslant t_2 \quad \text{(a)}$$

(ii) The probability of more than one change in $(t, t+\Delta t)$ is $O(\Delta t)$.

(iii) The probability of no change in the interval $(t, t+\Delta t)$ is:

$$1-\left\{\frac{-\alpha t_1(t-t_2)+\beta t_2(t-t_1)}{t(t_2-t_1)}\right\}\Delta t-O(\Delta t) \qquad \text{(b)} \quad (3.79)$$

in which the parameters α and β are related to the intensity as defined previously (see also reference[65]). It is convenient for a unified representation of the deformational behaviour, to introduce two new parameters that are associated with the intensity factor as well as the two limiting time instants t_1, t_2 in the following manner:

$$a = \frac{\lambda t_1 t_2}{t_2-t_1}, \qquad b = \frac{\lambda t_2}{t_2-t_1} \qquad (3.80)$$

These parameters may be regarded as physical characteristics of a given structured solid and lead to the elements of the transition matrix as follows:

$$q_{ij}(t) = \begin{cases} -b+\dfrac{a}{t} & \text{for } i = 0, 1, 2, \dots \\ b-\dfrac{a}{t} & \text{for } j = i+1 \\ 0 & \text{otherwise} \end{cases} \qquad (3.81)$$

By employing these relations the Kolmogorov differential equation for the transient state of the stochastic deformation can be written as:

$$\frac{\mathrm{d}\boldsymbol{P}_{ij}(t)}{\mathrm{d}t} = -b\boldsymbol{P}_{ij}(t)+\frac{a}{t}\boldsymbol{P}_{ij}(t)+b\boldsymbol{P}_{i,j-1}(t)-\frac{a}{t}\boldsymbol{P}_{i,j-1}(t),$$

$$t_1 \leqslant t \leqslant t_2 \tag{3.82}$$

the solution of this differential equation has been investigated in[65] and shows to be of the following form:

$$\boldsymbol{P}_{ij}(t) = \begin{cases} \dfrac{[b(t-t_1)-a\ln(t/t_1)]^{j-i}}{(j-i)!}t^a e^{-bt}, & j \geqslant i, \quad t_1 \leqslant t \leqslant t_2 \\ 0, & j < i \end{cases} \tag{3.83}$$

It is seen from the above developed theory of stochastic deformations, that with the exclusion of the deformational behaviour leading to fracture, the transition probability shows properties of invariance with respect to the purely elastic and the steady-state deformation process. More specifically for these two ranges of material behaviour, the transition probability satisfies the semi-group property. However, in the transient state the semi-group property is not satisfied and consequently the random process is of the non-Markovian type, but can be considered as a quasi-Poisson process. The significance of the intensity factor which has been introduced in the formulation is evident. This factor is of course determined by the condition of a given microstructure as well as the external boundary conditions imposed on the macroscopic material body.

Summarizing the important features of the random deformation process, it may be stated that such a process in a probability space $[U^\infty, \mathscr{F}^\infty, \mathscr{P}^u]$ is a sequence of random variables $\{\boldsymbol{u}(X; t)\}$ or $\boldsymbol{u}_t(X)$ each member of the sequence being generated from its previous one by an automorphism T_t, $t \in R^+$. The random process can be subdivided into three ranges according to whether the process represents the elastic, transient or steady-state deformations of the material and excluding the range of the microstructural changes leading to fracture. The distinct features of the different types of deformation can be described however in terms of two basic quantities, i.e. the transition matrix $\boldsymbol{Q}(t)$ and the transition probability $\boldsymbol{P}(t)$. Thus in terms of $\boldsymbol{Q}(t)$ it can be stated that:

$$Q(t) \rightarrow \begin{cases} q_{ii} = 0 \quad q_{ij} = 0, \quad j = i+1 & \text{(elastic deformations)} \\ q_{ii} = -\lambda \quad q_{ij} = \lambda, \quad j = i+1 & \text{(steady-state deformations)} \\ q_{ii} = \dfrac{-\lambda t_2(t-t_1)}{t(t_2-t_1)}, \quad q_{ij} = \dfrac{\lambda t_2(t-t_1)}{t(t_2-t_1)} \quad j = i+1 & \text{(transient)} \end{cases} \tag{3.84}$$

Further, in terms of the transition probability the distinction can be made as follows:

$$\begin{aligned} &\boldsymbol{P}_{ij}(t) = \delta_{ij} \qquad \boldsymbol{P}(t)\boldsymbol{P}(s) = \boldsymbol{P}(t+s), \qquad \text{(elastic)} \\ &\boldsymbol{P}_{ij}(t) = f(t) \qquad \boldsymbol{P}(t)\boldsymbol{P}(s) = \boldsymbol{P}(t+s), \qquad \text{(steady-state)} \\ &\boldsymbol{P}_{ij}(t) = g(t) \qquad \boldsymbol{P}(t)\boldsymbol{P}(s) \neq \boldsymbol{P}(t+s), \qquad \text{(transient)} \end{aligned} \tag{3.85}$$

in which the form of the functions $f(t)$ and $g(t)$ are given by equations (3.74) and (3.83), respectively.

It should be noted, that analogous considerations to that given for the deformation space concerning the change of the probability measure in that space, apply also to the stress space provided the process in the latter is again based on the assumption of the Markovian character of the process. Thus, for instance, for two Borel sets E_r and E_{r+1} in $\mathscr{F}^\sigma$ corresponding to time instants, t_r, t_{r+1} on the real line R^+ one can write for the probability distribution or the probability measure on Ξ the following relation:

$$\mathscr{P}^\sigma\{E_{r+1}, t_{r+1}\} = \boldsymbol{P}^\sigma\{t_{r+1}, t_r\}\mathscr{P}^\sigma\{E_r, t_r\} \tag{3.86}$$

If the process under consideration is of the homogeneous type, then analogously to (3.58) the following relation will hold:

$$\mathscr{P}^\sigma\{t\} = \mathscr{P}\{\sigma(t)\} = \boldsymbol{P}^\sigma(t)\mathscr{P}\{\sigma(0)\} = \boldsymbol{P}^\sigma(t)\mathscr{P}^\sigma\{0\} \tag{3.87}$$

Identifying the measure $\boldsymbol{P}^\sigma\{t\}$ with the transition probability of a Markov process leads to the following Kolmogorov differential equation:

$$\frac{d\boldsymbol{P}^\sigma(t)}{dt} = \boldsymbol{Q}^\sigma\boldsymbol{P}^\sigma(t), \quad \boldsymbol{P}^\sigma(0) = I \tag{3.88}$$

Although the assumption of the Markovian character of the random process in the stress space Ξ appears to be abstract, it leads nevertheless to results that are meaningful for physical processes (see reference[76]).

(C) *Some remarks on ergodic theorems*

With the development of a general theory of stochastic deformations from a measure theoretical and functional analytic point of view as shown above, it is perhaps appropriate to conclude this section with some remarks on the ergodic problem indicated earlier in Chapter I. Thus a brief discussion on some ergodic theorems of both classical and probabilistic systems is given below.

The ergodic approach in classical statistical mechanics is mainly concerned with establishing equilibrium values of the dynamic variables of a system. Thus, if $X[p(t), q(t)]$ is considered as a dynamic variable in terms of the generalized coordinates and momenta of a given system, the equilibrium value of X is given by the "time average" such that:

$$\overline{X} = \lim_{t\to\infty} \frac{1}{t}\int_0^t X[p(\tau), q(\tau)]\,\mathrm{d}\tau \tag{3.89}$$

One can define an invariant motion in the "phase-space $\Gamma(p, q)$" by the following relation:

$$H(p, q) = E \tag{3.90}$$

H being the Hamiltonian and E the energy of the system in its initial state, i.e. $E = H[p(0), q(0)]$.

If $\Omega \subset \Gamma$ is a region in the phase space compatible with the condition:

$$E - \Delta E < H(p, q) < E + \Delta E \tag{3.91}$$

the "phase average" is then defined by:

$$\langle X \rangle = \lim_{\Delta E \to 0} \frac{\int_\Omega X(p, q)\,\mathrm{d}p\,\mathrm{d}q}{\int_\Omega \mathrm{d}p\,\mathrm{d}q} \tag{3.92}$$

The ergodic theorem asserts that:

$$\overline{X} = \langle X \rangle \tag{3.93}$$

for all but a negligible class of initial states $[p(0), q(0)]$. This classical ergodic theorem is based on the mathematical structure of a classical dynamical system characterized by:

(i) a Γ-space which is the same as the R^{2n} Euclidean space,
(ii) a measure $\mu(p, q)$ defined on Γ,
(iii) a group of measure preserving automorphism T_t.

The above concept of a classical dynamical system can be extended to a measure space represented by the triplet $[X, \mathscr{F}, \mu]$ in which $\mathscr{F}$ is the σ-algebra of Borel sets in X and μ a measure on X. If T_t is a measure preserving automorphism, then:

$$\mu(T_t E) = \mu(E) \quad \text{for all } t \in R^+ \text{ and } E \in \mathscr{F} \tag{3.94}$$

However, instead of this classical system, one can also define an "abstract dynamical system" which is clearly representable by the quadruplet $[X, \mathscr{F}, \mu, T_t]$ (see Jancel[22]). The advantage of the concept of an abstract dynamical system over the classical system representation becomes apparent if one considers, for example, the Hamiltonian system. In that case X which is also identical with Γ is the real Euclidean space R^{2n}. The σ-algebra $\mathscr{F}$ may be thought of as open spheres in the R^{2n} space as given by relation (3.91) and μ as a Borel or Lebesgue measure in that space. The automorphism associated with the Hamiltonian equations then defines a continuous flow on the R^{2n} space permitting the consideration of the evolution of the system in contrast to the classical form, that presumes this motion to be a stationary one. In order to state ergodic theorems for systems represented by $[X, \mathscr{F}_t\, \mu, T_t]$ one may consider a function $f(x) \in X$, where the time average of f, if it exists, is expressed by:

$$\bar{f} = \lim_{N\to\infty} \frac{1}{N} \sum_{n=0}^{N-1} f(T_t^n x), \qquad n \in Z^+ \tag{3.95}$$

for a discrete system and

$$\bar{f} = \lim_{t\to\infty} \frac{1}{t} \int_0^t f(T_t x)\,\mathrm{d}t, \qquad t \in R^+ \tag{3.96}$$

for a continuous system. The space average of $f(x)$ can be written as:

$$\langle f \rangle = \int_X f(x)\,\mathrm{d}\mu \tag{3.97}$$

with the condition that:

$$\mu(X) = 1 \tag{3.98}$$

The ergodic theorems for abstract dynamical systems[77,78] are concerned with the existence of limits of time averages associated with certain classes of functions f defined on $[X, \mathscr{F}, \mu, T_t]$ and with the determination of the conditions under which these time averages become equal to the space averages. Some of the theorems are stated below:

T.1 (Birkhoff)[79]: Let $[X, \mathscr{F}, \mu]$ be a measure space and f a summable function on $L_1(X, \mu)$. If T_t is a measure preserving automorphism not necessarily invertible, then for any $\varepsilon > 0$

$$\left| \lim_{N\to\infty} \frac{1}{N} \sum_{n=0}^{N-1} f(T_t^n x) - \bar{f} \right| < \varepsilon, \quad n \in Z^+ \tag{3.99}$$

The limit $\bar{f}$ is integrable and invariant almost everywhere. If in addition $\mu(x) < \infty$, then

$$\int_X \bar{f} \mathrm{d}\mu(x) = \int_X f \mathrm{d}\mu(x) \tag{3.100}$$

The proof of the above theorem is based on another theorem called the "maximal ergodic theorem" given below:

T.2 (Yosida and Kakutani[80]): Let $[X, \mathscr{F}, \mu]$ be a measure space and f an integrable function on X. If $\Omega \subset X$ is a set of elements x such that at least one of the sums

$$f(x)+f(T_t x)+f(T_t^2 x)+ \ldots +f(T_t^n x)$$

is positive, then

$$\int_\Omega f(x) \mathrm{d}\mu(x) \geqslant 0 \tag{3.101}$$

These theorems establish the existence of "time limits" in general circumstances. The condition under which the time average becomes equal to the space average has been stated by the classical ergodic theorem. For an abstract dynamical system the condition is that $\bar{f}$ has to be invariant under the transformation T_t. In this case one can define two subsets of X of non-zero measure associated with the invariance property of $\bar{f}$ under T_t. Thus the necessary and sufficient condition for a system to be ergodic is that any invariant measurable set has the measure 0 or 1. In other words, that the system is "indecomposable". Hence for an abstract dynamical system which is also ergodic:

$$\bar{f} \overset{\text{a.e.}}{=} \langle f \rangle \tag{3.102}$$

It should be noted that the ergodic theorems for an abstract dynamical system involves the convergence of some measure (3.99) on the measure space $[X, \mathscr{F}, \mu]$. This convergence is directly related to the motion of a system approaching equilibrium. Thus the limitation of the classical ergodic theorem in its inability to distinguish between an equilibrium and non-equilibrium situation is eliminated. Another restriction of the classical ergodic theorem lies in that it cannot be utilized to explain the irreversibility of the system. In classical statistical mechanics this restriction is conventionally overcome by introducing a certain "randomness assumption" to account for the flow in the phase space. Similarly in the present theory such an assumption is also made, but it is postulated that the system behaviour is representable by a Markov type random process. This postulate is in fact not far from the notion of an abstract dynamical system as may be seen from the short discussion below.

Thus considering X to be the space of all state variables of the form ${}^{\alpha}\boldsymbol{v}(t)$, $\alpha = 1, \dots, N$ (ensemble of microelements) and $\mathscr{F}$ the σ-algebra of Borel sets, discussed earlier in this section, i.e. containing the sets

$${}^{\ni}E = \{{}^{\ni}\boldsymbol{v}(t) < {}^{\alpha}\boldsymbol{v}(t) < {}^{\ni}\boldsymbol{v}(t) + \Delta^{\ni}\boldsymbol{v}(t)\}, \qquad \ni \in Z^{+} \tag{3.103}$$

further μ to be identical to a probability measure on ${}^{\ni}E$, e.g. $\mathscr{P}\{{}^{\ni}E\}$ with $\mathscr{P}\{X\} = 1$, then $[X, \mathscr{F}, \mathscr{P}]$ defines a probability space. In correspondence with the automorphism T_t one can define a conditional probability measure $\boldsymbol{P}$ such that whenever

$$\begin{aligned} &T_t\colon\ E_r = E_{r+1} \\ &\boldsymbol{P}(t)\mathscr{P}\{E_r\} = \mathscr{P}\{E_{r+1}\} \end{aligned} \tag{3.104}$$

The quadruplet $[X, \mathscr{F}, \mathscr{P}, \boldsymbol{P}]$ then defines a Markov process, if the following properties are satisfied:

$$\begin{aligned} &\text{(i) } \forall t \in R^{+}, \quad \|\boldsymbol{P}(t)\| \leqslant 1 \\ &\text{(ii) } \forall s, t \in R^{+}, \quad \lim_{s \to t} [\boldsymbol{P}(t) - \boldsymbol{P}(s)]\mathscr{P}\{{}^{\ni}E\} = 0 \\ &\text{(iii) } \forall s, t \in R^{+}, \quad \boldsymbol{P}(t+s) = \boldsymbol{P}(t)\boldsymbol{P}(s) \end{aligned} \tag{3.105}$$

It is seen that the Markov process formulation $[X, \mathscr{F}, \mathscr{P}, \boldsymbol{P}]$ is quite analogous to the representation by means of an abstract dynamical system $[X, \mathscr{F}, \mu, T_t]$. Moreover, since there is a relationship between $\boldsymbol{P}$ and T_t and between $\mathscr{P}$ and μ, there is a one-to-one correspondence between $[X, \mathscr{F}, \mathscr{P}, \boldsymbol{P}]$ and $[X, \mathscr{F}, \mu, T_t]$. This suggests that one can formulate ergodic theorems for a Markov process similar to those for an abstract dynamical system. Hence

T.3: Let $[X, \mathscr{F}, \mathscr{P}, \boldsymbol{P}]$ be a Markov process and $\boldsymbol{v}(t)$ a summable function on $L_1(X, \mathscr{P})$. Then

$$\left| \lim_{N\to\infty} \frac{1}{N} \sum_{n=0}^{N-1} \boldsymbol{P}^n \boldsymbol{v}(t) - E(\boldsymbol{v}(t)|\Omega) \right| \to 0 \tag{3.106}$$

in the L_1 norm and for a subset $\Omega \subset X$.

Hence for an invariant measure $\mathscr{P}$, one obtains:

$$\int \left| \lim_{N\to\infty} \frac{1}{N} \sum_{n=0}^{N-1} \boldsymbol{P}^n \boldsymbol{v}(t) - E(\boldsymbol{v}(t)|\Omega) \right| \mathrm{d}\mathscr{P} = 0 \tag{3.107}$$

It is seen that the above theorem is similar to T.1 (Birkhoff[79]) for an abstract dynamical system. In this context another theorem can be stated referred to as the "Hopf maximal ergodic lemma"[81], i.e.:

T.4: Let $\boldsymbol{v}(t) \in L_1(X, \mathscr{P})$ and define

$$\Omega = \{\boldsymbol{v}(t) \colon \sup_n(\boldsymbol{v}(t) + \boldsymbol{v}(t)\boldsymbol{P} + \dots + \boldsymbol{v}(t)\boldsymbol{P}^n) > 0\} \tag{3.108}$$

then:

$$\int_\Omega \boldsymbol{v}(t)\mathrm{d}\mathscr{P} \geqslant 0 \tag{3.109}$$

The proof of this theorem is given by Foguel[78]. As a consequence of the above theorems for Markov processes, it may be noted that by definition (3.105) the transition probability of the process satisfies the semi-group property. Hence the irreversibility of the system is "built-in". Further, from T.3 (3.106) it is readily seen that the transition probability has a convergence property, where the limit of convergence $E(\boldsymbol{v}(t)|\Omega)$ is in fact a conditional expectation of $\boldsymbol{v}(t)$ on a subset $\Omega \subset X$. Thus, in this sense the ergodic theorem for a Markov process forms a basis for calculating the conditional expectation of a random variable or a stochastic process, if the transition probability is known. Furthermore, if the subset Ω satisfies equation (3.108), then according to T.4, the average value of the stochastic process can be calculated.

So far as the applicability of the above theorems to the general theory of stochastic deformations is concerned, one can consider a subset $U \subset X$ and the corresponding measure space $[U, \mathscr{F}^u, \mathscr{P}^u]$ as a starting point. Since the latter has been referred to as the "deformation space", the Markov process associated with this space can be denoted by $[U, \mathscr{F}^u, \mathscr{P}^u, \boldsymbol{P}^u]$. Thus equation (3.106) in this case can be written as:

$$\int \left| \lim_{N\to\infty} \frac{1}{N} \sum_{n=0}^{N-1} (\boldsymbol{P}^u)^n \boldsymbol{u}(t) - E(\boldsymbol{u}(t)|U_s) \right| \mathrm{d}\mathscr{P}^u = 0 \tag{3.110}$$

in which $\boldsymbol{u}(t) \in U$ is contained in $\boldsymbol{v}(t)$ and U_s is a subset of U. Thus the ergodic theory for the case of a general theory of stochastic deformations simply establishes a method of evaluating the average value of the deformation process. For the steady-state deformations one could consider that:

$$\int \left| \lim_{N\to\infty} \frac{1}{N} \sum_{n=0}^{N-1} (\boldsymbol{P}^u)^n \boldsymbol{u}(t) - E(\boldsymbol{u}(t)|U_s) \right| \mathrm{d}\mathscr{P}^u < \varepsilon \tag{3.111}$$

for any $\varepsilon > 0$.

In conclusion of the brief discussion on ergodic theorems, it may be noted that according to relation (3.93) or (3.95) it has to be ensured that t is large enough that the dynamical system $[X, \mathscr{F}, \mathscr{P}, \boldsymbol{P}]$ or $[U, \mathscr{F}^u, \mathscr{P}^u, \boldsymbol{P}^u]$ passes through all possible states before attaining equilibrium. Whilst this may be the case for a gas or fluid, it will not be so for a structured solid. Hence, it may be assumed that "metric transitivity" or ergodicity in the case of a general theory of stochastic deformation of a structured solid cannot be referred to the complete σ-algebra $\mathscr{F}^u$ but rather to a subset $\mathscr{F}^u_s$ of $\mathscr{F}^u$.

Since the Markovian approach to the deformation process deals from the onset with an irreversible phenomenon in which the evolution with time of a state variable is more significant than its time average, considerations of the ergodic theorems within the framework of the present theory appear to be unnecessary.

3.4 Material operators in micromechanics

The significance of a material functional or material operator in the micromechanics of structured solids has been frequently mentioned in the foregoing chapters. In particular, it has been pointed out in the preceding sections, that the material operator provides a link between the stress and deformation space. The subsequent discussion on material operators is concentrated on clarifying this concept further and its application in the formulation of response relations of a structured medium. There are two types of material operators, i.e. one which is used in the response relations of individual structural elements and which includes interaction effects between the elements and another pertaining to the response relations of an ensemble of microelements in a given mesodomain.

In accordance with the basic concepts of probabilistic micromechanics that have been stated in Chapter 1, these material operators or functionals permit to replace the conventional relations referred to as constitutive laws. Material operators, in general, contain those stochastic variables or functions of such variables that are representative of a specific structured medium and which can be constructed from a knowledge of the variables or their distributions on basis of experimental data.

Hence, a material operator as seen in the present theory is aimed at realistically representing the material characteristics of a given medium. The material constants used in most theories of continuum mechanics have shown to be completely inadequate in considerations of structured media, since many effects influence the behaviour of such materials, in particular, those related to the microstructure.

(A) *Material operator for the microelement*

From a mathematical point of view the material operator for a structural or microelement of the medium is a mapping between the stress and deformation space such that:

$$M: \Xi \rightarrow U$$

and (3.112)

$$M^{-1}: U \rightarrow \Xi$$

in which the microelement operator M may be non-linear in general. Considering that ${}^{\alpha}\boldsymbol{\sigma} \in \Xi$ and ${}^{\alpha}\boldsymbol{u} \in U$ as discussed in the previous section, for a domain $\mathscr{D} \subset \Xi$ and a corresponding range $\mathscr{R} \subset U$ one can write:

$$ {}^{\alpha}\boldsymbol{u}(t) = M({}^{\alpha}\boldsymbol{\sigma}(t)) \qquad (3.113) $$

It should be noted that the M operator contains random variables or functions of such variables and hence within the framework of a general deformation theory in order to formulate "macroscopic" response relations from the "microscopic" ones, the distribution of M will have to be taken into account. With reference to earlier publications[82,83] the operator M contains two "transform operators". Thus the transform operator ${}^{\alpha}A$ designates the relation between the "input" or microstress to the "output" in terms of deformations or strains, if the microelement for simplicity is regarded as homogeneous and another transform operator ${}^{\alpha\beta}B$ relating the surface stress input to the response within the boundary zone between two microelements. Hence

$$ {}^{\alpha}A\colon {}^{\alpha}\Xi \rightarrow {}^{\alpha}U, \qquad {}^{\alpha\beta}B\colon {}^{\alpha\beta}\Xi \rightarrow {}^{\alpha\beta}U \qquad (3.114) $$

so that an implicit form of the microelement operator can be written as:

$$ M = M({}^{\alpha}A, {}^{\alpha\beta}B) \qquad (3.115) $$

If the transform operators are linear ones, the following system relations can be written:

$$ \begin{aligned} {}^{\alpha}\boldsymbol{w}(t) &= {}^{\alpha}\boldsymbol{A}(t){}^{\alpha}\boldsymbol{\xi}(t) \\ {}^{\alpha\beta}\boldsymbol{d}(t) &= {}^{\alpha\beta}\boldsymbol{B}(t){}^{\alpha\beta}\boldsymbol{\xi}(t) \end{aligned} \qquad (3.116) $$

Then M becomes a function of the two transform operators such that:

$$ M = M({}^{\alpha}\boldsymbol{A}(t), {}^{\alpha\beta}\boldsymbol{B}(t)) \qquad (3.117) $$

The transform operators, as shown later, are either integral operators or if the inverse form is used, differential operators. Analogously to continuum mechanics where often from an assessed strain field the corresponding stress field is determined by the use of constitutive relation, this can be done in probabilistic micromechanics also, but involves the inverted operator, i.e. M^{-1}. Hence the problem of the invertibility of the material operator in general is of utmost importance. Investigations to this effect (see also reference[65]) have lead to the following theorem:

T.1: M is strictly monotone for some dense subsets Ξ_d and U_d of Ξ and U, respectively.

Hence by writing

$${}^{\alpha}\boldsymbol{u}(t) = M({}^{\alpha}\boldsymbol{\sigma}(t))$$

for ${}^{\alpha}\boldsymbol{u} \in U_d$ and ${}^{\alpha}\boldsymbol{\sigma} \in \Xi_d$, it can be shown that a non-degenerate bilinear form $\langle \cdot , \cdot \rangle$ between the elements of U_d and Ξ_d with respect to the operator M can always be found (Tonti[84]) such that:

$$\langle \{M({}^{\ni}\boldsymbol{\sigma}) - M({}^{\text{я}}\boldsymbol{\sigma})\}, \; \{{}^{\ni}\boldsymbol{\sigma} - {}^{\text{я}}\boldsymbol{\sigma}\}\rangle > 0 \tag{3.118}$$

in which ${}^{\ni}\boldsymbol{\sigma}$, ${}^{\text{я}}\boldsymbol{\sigma}$ are specific values of the microstress variable ${}^{\alpha}\boldsymbol{\sigma}$ within the dense set of Ξ_d. Thus the operator M is "strictly monotone". It has been shown in reference[65] that for such dense subsets Ξ_d and U_d, the operator M is invertible.

An extension of this theorem is even more important, since considerations of an ensemble of microelements and their response relations in the general deformation theory is required. In this context, it can be shown that Ξ_d and U_d converge to the spaces Ξ and U, respectively, so that the subspaces of stress and deformation can be thought of as limit sets of Ξ_d and U_d whereby the material operator is "strongly monotone". The second criterion, i.e. the invertibility of the M operator on the basis of "strong monotonicity" is more desirable, since it ensures the invertibility at all times. It is evident, that for a certain class of structured media the response behaviour by consideration of the associated function spaces is such that the latter are dense enough so that almost always $\Xi_d = \Xi$ and $U_d = U$. In this case the condition of "strong monotonicity" becomes that of "strict monotonicity". However, a wider class of materials will be representable in their response behaviour by the weaker condition, i.e. that of strong monotonicity of the material operator and that condition may be sufficient for the proof of its invertibility. Thus considering that M is invertible, then the converse relation to (3.113) becomes:

$${}^{\alpha}\boldsymbol{\sigma}(t) = M^{-1}({}^{\alpha}\boldsymbol{u}(t)) \tag{3.119}$$

The inverse operator M^{-1} of a microelement in terms of the transform operators ${}^{\alpha}A$, ${}^{\alpha\beta}B$ or their inverses can be expressed symbolically by:

$$M^{-1}(t) = M^{-1}\{{}^{\alpha}A^{-1}(t), {}^{\alpha\beta}B^{-1}(t)\} \tag{3.120}$$

For the linear case, the above form corresponding to (3.117) becomes then:

$$M^{-1}(t) = M^{-1}\{{}^{\alpha}A^{-1}(t), {}^{\alpha\beta}\boldsymbol{B}^{-1}(t)\} \tag{3.121}$$

Further details of the material operator M will be given in Chapter IV concerned with the application of the general theory of stochastic deformation, and where the explicit expressions of the operators ${}^{\alpha}\boldsymbol{A}(t)$ and ${}^{\alpha\beta}\boldsymbol{B}(t)$ for two cases are discussed.

(B) *Material operator for the mesodomain*

In the above paragraph the material operator for a microelement has been considered. It is now possible to formulate such an operator for the mesodomain, which will be denoted by $\mathscr{M}(t)$. Since, by postulation, the material operator $M(t)$ is a function of several mechanical and physical random parameters that are significant for the response of a structured medium, it can be regarded as a stochastic variable and a probability measure can be assigned to it. Thus $\mathscr{M}(t)$ for a particular mesodomain will be determined by the average of the microelement operator M taken with respect to its distribution. This can be expressed by:

$$\mathscr{M}(t) = \sum M(t)\,\Delta\mathscr{P}\{M(t)\} = \sum M(t)p\{M(t)\} \tag{3.122}$$

where $\Delta\mathscr{P}\{M(t)\}$ or $p\{M(t)\}$ is the probability density of the microelement operator $M(t)$. Since the latter is also a function of two transform operators, the probability distribution may be written explicitly as follows:

$$\mathscr{P}\{M(t)\} = f[\mathscr{P}\{{}^{\alpha}A(t), {}^{\alpha\beta}B(t)\}] \tag{3.123}$$

As stated previously, the material operator $M(t)$ is regarded as a stochastic variable and hence its change and the change of its distribution with time is of interest. With reference to the statement given in Section 3.3 (ii) dealing with the probability distribution of microstresses which depends on the distribution of the operator

$M(t)$, relation (3.119) must now be interpreted as an inverse mapping $M^{-1}(t)$ from the deformation to the stress space. Equivalently, as stated in Section 3.3 (iii) the probability measure on the stress space is linked to that on the deformation space by the material operator or more precisely by its measure. Thus the distribution of the microstress ${}^{\alpha}\sigma(t)$ may be visualized as some kind of product measure derived from the probability distributions of the material operator $M(t)$ and that of microdeformations.

In particular, considering the inverse operators and microdeformations as statistically independent random variables, the following relation between the distributions will exist:

$$\mathscr{P}\{\sigma(t)\} = \mathscr{P}\{\boldsymbol{M}^{-1}(t)\}\mathscr{P}\{\boldsymbol{u}(t)\} \tag{3.124}$$

The concepts of statistical independence and of the product measure together with the assumptions of the deformation processes in the deformation and stress space to be of a Markovian character lead in the present theory to differential equations that represent the evolution of the system in these spaces. It involves however the distribution of the inverse material operator, $\mathscr{P}\{\boldsymbol{M}^{-1}(t)\}$. The solutions of these differential equations yield then insight into the probability distribution of the microscopic material operator M at any time during the deformation process. However, in general such a procedure may not hold if, for instance, the probabilistic and topological structure of the stress space is unknown. In that case by means of relation (3.124) and a linearity assumption of the inverse operator, the construction of a non-degenerate bilinear form with respect to the inverse operator is still possible, which then includes a proper topological structure on the stress space Ξ (see also Moreau[67] and Tonti[68]).

3.5 Governing equations and response relations of structured solids

In continuum mechanics the relevant "field quantities" associated with the deformational behaviour of a homogeneous medium are specified in terms of "point functions". The resulting mathematical relations are then considered as "field equations". Such equations together with the constitutive laws and stress equations of motion are unified in order to describe the motion of a material point in the continuum. Thus the outstanding characteristic in continuum theories is the consideration of point functions. This notion is however of no significance in probabilistic micromechanics, since this theory considers the field quantities as stochastic variables or functions of such variables, which by definition are "set functions". Furthermore, due to the basic assumption that the deformation process is a stochastic process of a special type, the corresponding formulation is mainly concerned with the evolution of probability distributions or measures and the resulting differential equations may be referred to as "governing equations of a structured medium". More precisely the governing equations that represent the general deformations consist of a "set of differential equations" which involve probability measures of the relevant stochastic variables. Referring to relations (3.58) and (3.87) of Section 3.3 of this chapter, it is readily seen that the following relation can be established:

$$\boldsymbol{P}^{\sigma}(t)\mathscr{P}\{\boldsymbol{\sigma}(0)\} = \mathscr{P}\{\boldsymbol{M}^{-1}(t)\}\boldsymbol{P}^{u}(t)\mathscr{P}\{\boldsymbol{u}(0)\} \tag{3.125}$$

involving the distribution of the material operator. In order that equation (3.124) be satisfied at all instants of time t, one obtains for $t = 0$ the following relation:

$$\mathscr{P}\{\boldsymbol{\sigma}(0)\} = \mathscr{P}\{\boldsymbol{M}^{-1}(0)\}\mathscr{P}\{\boldsymbol{u}(0)\} \tag{3.126}$$

which together with equation (3.125) yields:

$$\boldsymbol{P}^{\sigma}(t)\mathscr{P}\{\boldsymbol{M}^{-1}(0)\}\mathscr{P}\{\boldsymbol{u}(0)\} = \mathscr{P}\{\boldsymbol{M}^{-1}(t)\}\boldsymbol{P}^{u}(t)\mathscr{P}\{\boldsymbol{u}(0)\} \tag{3.127}$$

Thus, if the initial distribution $\mathscr{P}\{\boldsymbol{u}(0)\}$ is specified in accordance with experimental observations as will be discussed in Chapter V and has a non-zero value, it may be concluded from (3.127) that:

$$\boldsymbol{P}^{\sigma}(t)\mathscr{P}\{\boldsymbol{M}^{-1}(0)\} = \mathscr{P}\{\boldsymbol{M}^{-1}(t)\}\boldsymbol{P}^{u}(t) \tag{3.128}$$

It is seen that this equation relates the probability distribution of the material operator at any time t to that at $t = 0$, in terms of the transition probabilities $\boldsymbol{P}^{\sigma}(t)$ and $\boldsymbol{P}^{u}(t)$.

In the terminology of the present theory relation (3.128) can be regarded as one form of the governing equations of material operators by means of $\boldsymbol{P}^{u}$ and $\boldsymbol{P}^{\sigma}$. To illustrate this remark consider the steady-state deformation process and the one-step transition probability $\boldsymbol{P}^{u}(t)$ as discussed earlier (eqn. (3.76)). If in this case the time instant t_2 is considered as the initial one, i.e. if $t_2 = 0$ and $\boldsymbol{P}^{\sigma}(t_2)$ is equal to $\boldsymbol{I}$, it follows that:

$$\mathscr{P}\{\boldsymbol{M}^{-1}(t)\} = e^{\lambda t}\mathscr{P}\{\boldsymbol{M}^{-1}(0)\} \tag{3.129}$$

However, for the more general case, e.g. when the two transition probabilities are arbitrary functions of time, one can differentiate expression (3.128) to give:

$$\mathscr{P}\{\boldsymbol{M}^{-1}(0)\}\frac{\mathrm{d}\boldsymbol{P}^{\sigma}(t)}{\mathrm{d}t} = \frac{\mathrm{d}\mathscr{P}\{\boldsymbol{M}^{-1}(t)\}}{\mathrm{d}t}\boldsymbol{P}^{u}(t) + $$

$$+\mathscr{P}\{\boldsymbol{M}^{-1}(t)\}\frac{\mathrm{d}\boldsymbol{P}^{u}(t)}{\mathrm{d}t} \tag{3.130}$$

This relation is quite general since it includes all the possible stages of the deformational behaviour of a structured medium, with the exclusion of breakdown of the microstructure leading to fracture. If one considers elastic deformations only, the above relation is seen to reduce to:

$$\frac{\mathrm{d}\mathscr{P}\{\boldsymbol{M}^{-1}(t)\}}{\mathrm{d}t} = 0 \tag{3.131}$$

since both $\boldsymbol{P}^{u}(t)$ and $\boldsymbol{P}^{\sigma}(t)$ are constant. This result is apparent since during an elastic deformation the material properties in general can be regarded to remain constant. On the other hand, it can be shown that for steady-state deformations, the relation (3.130) leads to:

$$\frac{\mathrm{d}\mathscr{P}\{\boldsymbol{M}^{-1}(t)\}}{\mathrm{d}t} - \lambda\mathscr{P}\{\boldsymbol{M}^{-1}(t)\} = 0 \tag{3.132}$$

which has a solution given by (3.129).

Thus it is seen that relation (3.130) can be looked upon as the key governing equation of a structured medium. If the transition probabilities $\boldsymbol{P}^{u}(t)$ and $\boldsymbol{P}^{\sigma}(t)$ are given in terms of the Kolmogorov relations, i.e. (3.65) and (3.88), the expression (3.130) can be written as follows:

$$\mathscr{P}\{\boldsymbol{M}^{-1}(0)\}\,\boldsymbol{Q}^{\sigma}(t)\boldsymbol{P}^{\sigma}(t) = \frac{\mathrm{d}\mathscr{P}\{\boldsymbol{M}^{-1}(t)\}}{\mathrm{d}t}\boldsymbol{P}^{u}(t) + $$

$$+\mathscr{P}\{\boldsymbol{M}^{-1}(t)\}\,\boldsymbol{Q}^{u}(t)\boldsymbol{P}^{u}(t) \tag{3.133}$$

and by virtue of equation (3.128) the above relation reduces to:

$$\frac{d\mathscr{P}\{M^{-1}(t)\}}{dt} + (Q^{u} - Q^{\sigma})\mathscr{P}\{M^{-1}(t)\} = 0 \tag{3.134}$$

If for simplification of the analysis the transition matrices Q^{u} and Q^{σ} are considered to be time-independent, the solution of the above differential equation is given by:

$$\mathscr{P}\{M^{-1}(t)\} = \mathscr{P}\{M^{-1}(0)\}\exp[-(Q^{u} - Q^{\sigma})t] \tag{3.135}$$

from which one obtains the following relation between the two transition matrices:

$$Q^{\sigma} = Q^{u} + \frac{1}{t}[\ln \mathscr{P}\{M^{-1}(t)\} - \ln \mathscr{P}\{M^{-1}(0)\}] \tag{3.136}$$

It should be noted that the above result is more rigorous than that presented in earlier publications[76,85], since in this formulation the distribution of the material operator is involved, whereas previously this distribution has not been considered. This result has an interesting consequence in that it permits the determination of the stress transition matrix from the knowledge of the material operator distribution and of the deformation transition matrix. The latter distributions are amenable to experimental investigations. For the time-dependent case, one can analogously obtain a solution of (3.134) in the form of:

$$\int_0^t Q^{\sigma}(\tau)d\tau = \int_0^t Q^{u}(\tau)d\tau + $$
$$+ [\ln \mathscr{P}\{M^{-1}(t)\} - \ln \mathscr{P}\{M^{-1}(0)\}] \tag{3.137}$$

In conclusion, it may be stated that, if the probabilistic structure of the deformation and stress space are well defined, then by an appropriate choice of the stochastic process representing the deformation process for a given medium as a whole or parts of it, the evolution of the probability measures on these spaces can be formulated. Since these two spaces are related to each other by the material operator that itself contains stochastic parameters, the evolution of the probability measure on U and Ξ is coupled to that of the material operator. This view has been expressed in the present analysis in formulating the key governing equations for a structured medium.

Finally, it may become necessary in certain problems to include "body forces" in the deformation analysis. In this case it is possible to use the operational representation of the deformational behaviour of a structured solid as presented in this chapter and that part of the material operator M which permits the inclusion of body forces on the same scale as the effect arising from a bonding potential. It should also be mentioned that in the present formulation, no reference has been made to the thermodynamics of the deformation process. Whilst such a study is of considerable interest its inclusion lies outside the scope of this monograph. However, studies are at present being carried out to consider the deformation analysis from a thermodynamics point of view and in particular to relate the "transient" state of deformation to a "stability criterion" concerning the relevant stochastic parameters in terms of information theory and information stability of random processes[86].

IV. Applications of the Probabilistic Micromechanics Theory

4.1 Introduction

In accordance with earlier statements that two classes of structured solids are of particular interest in our investigations, the application of the general theory of stochastic deformation to these groups of materials will be considered in this chapter. Thus, first the elastic response behaviour of polycrystalline solids and then the mechanical response of a two-dimensional fibrous system will be treated. Numerical considerations in these sections are based on available experimental data for such materials. The last section of this chapter will consider the important case of the mechanical relaxation of crystalline solids from the point of view of the theory presented in the foregoing chapters.

4.2 Microstructural properties and the elastic response of polycrystalline solids

Polycrystalline solids such as metals comprise of crystals that are separated from each other by "boundaries". Each crystal is a three-dimensional network of atoms arranged in unit cells possessing an order which tends to reduce their potential energy to a minimum. However, the lattice of actual crystals is never perfect due to the presence of vacancies, stacking faults, line imperfections such as dislocations, etc. Furthermore, impurity atoms frequently penetrate the lattice structure causing other irregularities. Apart from the geometrical defects, lattice vibrations (see references[87,88]) contribute to the discrepancy between the observed mechanical response characteristic of a single crystal and that obtained theoretically on the assumption that the lattice is perfect.

It is evident that the effects of such imperfections and, in particular, those due to the presence of grain boundaries should be included in the response formulation of such materials. For this purpose the two main effects, i.e. dislocations and grain boundary effects will be included in the subsequent analysis of the elastic behaviour of polycrystalline solids. In the sense of the present theory a typical "microelement or grain" of a polycrystalline solid is indicated in Fig. 17(a, b). Figure 17(a) represents a silicon steel transmission electron microscope photograph under a magnification of 5700×. Figure 17(b) shows schematically the position of a typical crystal of the material and the external and body coordinate frames. In addition a point on the grain boundary of the αth crystal is indicated together with the corresponding surface coordinates that will be employed later. The configuration shown is that for the undeformed state of the solid as already given previously in dealing with the deformation kinematics (see Fig. 13, Chapter II) of such materials.

In order to deal with the two effects significant for the response behaviour of a polycrystalline solid it is convenient to treat them separately. As mentioned in the foregoing chapter the response behaviour will be presented in an operational form involving the material operator of individual microelements or grains and the operator associated with the grain boundary behaviour. The relation between the internal stress and deformation in terms of these operators has been given earlier in (3.116) and (3.119) respectively, as well as the expression relating the relative deformation in the grain boundary zone to the stress acting in that zone. The operators have been designated by ${}^{\alpha}\boldsymbol{A}(t)$ and ${}^{\alpha\beta}\boldsymbol{B}(t)$. In general, these operators or their inverse forms will be as follows:

$$ {}^{\alpha}\boldsymbol{A}^{-1}(t) = {}^{\alpha}\boldsymbol{A}^{-1}[\boldsymbol{E}, \boldsymbol{\Gamma}, G, \nu, \varrho_d, l, f_d, t, \nabla] \tag{4.1} $$

in which, as shown subsequently, the parameters in the argument of ${}^{\alpha}\boldsymbol{A}^{-1}(t)$ are the elastic tensor modulus $\boldsymbol{E}$ of the crystal, $\boldsymbol{\Gamma}$ the crystallographic orientations of the crystal structure, G the shear modulus, ν Poisson's ratio, ϱ_d the dislocation density, l a "characteristic length" of the inactivated Frank–Read source, f_d the fraction of mobile dis-

(a) Silicon steel TEM 5700×

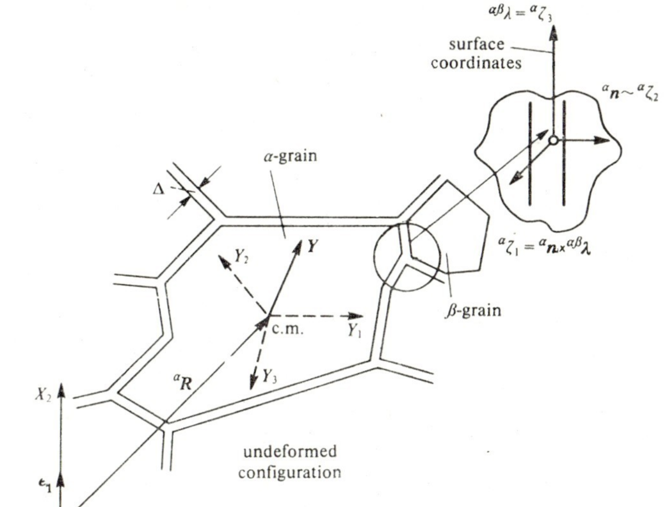

(b) Schematic for polycrystalline solid
Fig. 17. Polycrystalline materials.

locations, t the time and $\boldsymbol{\nabla}$ a gradient operator on the internal deformations. These parameters and an explicit form of the operator ${}^{\alpha}\boldsymbol{A}^{-1}(t)$ will be discussed below on the basis of a dislocation model proposed in earlier investigations[89,90]. One can express similarly the inverse interaction operator ${}^{\alpha\beta}\boldsymbol{B}^{-1}(t)$ relating the stress and deformation occurring within the boundary zone in general as follows:

$$ {}^{\alpha\beta}\boldsymbol{B}^{-1}(t) = {}^{\alpha\beta}\boldsymbol{B}^{-1}[\Psi_0, b, {}^{\alpha\beta}\boldsymbol{d}, t] \tag{4.2} $$

where Ψ_0 designates an equilibrium value of the interfacial potential, b a material characteristic constant and ${}^{\alpha\beta}\boldsymbol{d}$ the small relative displacement between "coincidence cell" points of one crystal with respect to its contiguous neighbour as indicated earlier in (2.23) (Chapter II). In order to obtain explicit forms of the operators the effects will now be discussed.

(A) *Dislocation effects*

The various dislocation mechanisms and their effect on the behaviour of polycrystalline solids in general are well understood. Several theories concerning the motion of dislocations in crystals are available and the reader is referred to the work of Cottrell[91], Kröner[92,93], Hirth and Lothe[94], Zorski[95], and others[96,97] mentioned in the bibliography of this text. In general, dislocations in a polycrystalline solid form a three-dimensional network that is randomly distributed in each crystal or microelement. Segments of the network are anchored to impurity atoms by the so-called Cottrell force. Upon the application of an external stress field to the microelement containing such a network of dislocations, a dislocation of a certain length starts to deform along a "slip-plane" in a sequence that is proportional to the applied stress. This topology was first introduced by Frank and Read[98] and is known as the Frank–Read source. This concept of dislocation motion is of utmost significance for the irreversible response of polycrystalline solids, since it forms the source for the occurrence of an extremely large number of dislocations in the crystal. For the purely elastic response however, it is possible to use a simplified model, which has been proposed in earlier investigations[89,90] on the basis of the following assumptions:

(a) The crystal matrix or the microelement is considered for simplicity of the analysis to exhibit a linear, isotropic elastic response behaviour in spite of the fact that it may contain imperfections as in an actual material.

(b) The density of the line dislocations per unit length in a unit volume of the αth crystal or microelement is inversely proportional to its average radius ${}^{\alpha}\bar{r}$. This assumption is based on the notion that, in general, a crystal prefers to grow within a closed packed structure such that the surface potential energy attains a minimum value[99]. From a geometrical point of view, by using a misfit area ${}^{\alpha}s$ and the average radius ${}^{\alpha}\bar{r}$ corresponding to an assumed spherical growth tendency, it is possible to define a simple form for the dislocation density[100], i.e.:

$$\varrho_d = \frac{6}{|\boldsymbol{b}|\,{}^{\alpha}\bar{r}} \tag{4.3}$$

in which $\boldsymbol{b}$ is the Burger vector. For a more accurate form see for instance[92].

(c) It is further assumed for simplification of the analysis, that the dislocation density does not vary from crystal to crystal. This of course will not be the case for real materials and also neglects the effects of possible fluctuations in the line dislocation density during the elastic response of the solid.

(d) A certain ratio of the line dislocations within the crystal are assumed to be "mobile".

(e) Whilst there are numerous possible slip systems in each crystal in the simple model suggested in earlier publications[89,90] only one slip direction ${}^{\alpha}O_{\beta j}$ ($\beta = 2, 3$) has been admitted for the motion of mobile dislocations, where ${}^{\alpha}O_{\beta j}$ denotes the direction cosines between the attached body frame of the microelement and the fixed external frame. Thus ${}^{\alpha}y_2$ is the direction of the "bow-out" of the inactivated Frank–Read source, whilst ${}^{\alpha}y_3$ denotes the normal to the slip plane. This restriction on the model can however be lifted, as discussed below.

(f) The orientations $\boldsymbol{O}$ have a random distribution in a given mesodomain.

(g) The dislocations interact only with the externally applied stress field, but not amongst themselves.

Hence, in accordance with these assumptions (a)–(e) and in particular using the simplified model suggested in (e), the elastic fourth order isotropic tensor or elastic modulus for the crystal becomes:

$$^{\alpha}\boldsymbol{E} = \frac{2G\nu}{(1-2\nu)}\delta_{ij}\,\delta_{kl}+2G\delta_{ik}\,\delta_{jl} = {}^{\alpha}E_{ijkl} \tag{4.4}$$

and the contribution of dislocations on the basis of this model to the response behaviour of the single microelement can be expressed by:

$$\begin{aligned} \eta^{\alpha}\boldsymbol{\Gamma} &= \left[\frac{2}{3\pi}G(1-\nu)l^2 f_d \varrho_d\right]{}^{\alpha}O_{3i}{}^{\alpha}O_{2j}{}^{\alpha}O_{3k}{}^{\alpha}O_{2l} \\ {}^{\alpha}\boldsymbol{\Gamma} &= {}^{\alpha}O_{3i}{}^{\alpha}O_{2j}{}^{\alpha}O_{3k}{}^{\alpha}O_{2l} \end{aligned} \tag{4.5}$$

in which the symbols have the same meaning as stated earlier and the quantity η equals the bracketed term in (4.5). It should be noted that the quantity $^{\alpha}\boldsymbol{\Gamma}$ in equation (4.5) is not symmetric in the indices i, j, k and l due to the chosen model discussed in detail in reference.[89] The physical meaning of η is that it represents the work done by a force per unit length $|^{\alpha}\boldsymbol{F}|$ acting on a dislocation segment of the characteristic length l displacing it for a certain distance $^{\alpha}y_2$ times the number of dislocation segments that are mobile within the crystal. The transformation of the "local strain components" from the body frame to the external one introduces thereby the fourth order tensor $^{\alpha}\boldsymbol{\Gamma}$. By combining relations (4.4) and (4.5) the operator relating the deformations to the internal microstress $^{\alpha}\xi$ in the crystal for the simplified model becomes:

$$\begin{aligned} {}^{\alpha}A^{-1} &= \left[\frac{2G\nu}{(1-2\nu)}\delta_{ij}\,\delta_{kl}+2G\delta_{ik}\,\delta_{jl}+\right. \\ &\left.+\frac{2}{3\pi}G(1-\nu)l^2 f_d \varrho_d{}^{\alpha}O_{3i}{}^{\alpha}O_{2j}{}^{\alpha}O_{3k}{}^{\alpha}O_{2l}\right]\nabla_l \end{aligned} \tag{4.6}$$

or in a reduced form:

$$^{\alpha}A^{-1} = [^{\alpha}\boldsymbol{E}+\eta^{\alpha}\boldsymbol{\Gamma}]\cdot\nabla \tag{4.7}$$

By lifting the restriction imposed on the model by assumption (e), which permits the motion of dislocations on slip planes in one direction only, to the more general case of slip to occur on a multitude of planes and directions, requires the introduction of the distribution of orientations. Following the detailed study of this case as given in reference[101] the operator ${}^{\alpha}A^{-1}$ will take the following form:

$$ {}^{\alpha}A^{-1}\colon {}^{\alpha}A^{-1}_{ijk} = \left[\frac{2G^2}{G+4\eta}\,\delta_{ik}\,\delta_{jl} - \right. $$

$$ \left. - \frac{2G[3G\nu+4\eta(1+\nu)]}{3(1-2\nu)\,(G+4\eta)}\,\delta_{ij}\,\delta_{kl} \right] \nabla_l \qquad (4.8) $$

The above relation of the operator accounting for the effect of dislocations in a single microelement will be used subsequently to establish the material operator M. The latter will also include grain boundary effects as discussed below.

(B) *Grain boundary effects*

In order to introduce the effect of grain boundaries on the overall response behaviour of crystalline solids, it is necessary first to remark on the grain boundary topology. As already indicated in Chapter II of this text, the present considerations are based on the geometrical theory of Bollmann[36]. The latter theory represents an idealization of two crystals in terms of interpenetrating mathematical translation lattices. If these lattices involve low misfit angles, they are referred to as "O-lattices", whilst for high misfit angles they are called "O_2-lattices". The theory interprets in both these cases the lattices as the sum of all positions of "best fit" and are considered to represent the description of all possible boundaries between two given crystals. However, this theory does not supply quantitative values of the grain boundary energy. In the O-lattice theory the lattice of the αth crystal is assumed to be fixed, whilst the lattice of the βth crystal undergoes changes such as translation, rotation, expansion, etc. Since the relative orientation within a certain domain of the microstructure can be specified, the lattice of the βth crystal translates for instance in such a manner that one of its points coincides with one of the αth crystal lattice. Such a point is called "the lattice coincidence site". Due to the periodicity of the two lattices a "coincidence site lattice" can always be formed. A systematic study of this situation can be found in the work of Bollmann[36]. From considerations of the O-lattice structure one can establish a relation between two contiguous crystals in the following manner[102]:

$$ X^{(\beta L)} = {}^{\alpha\beta}T \cdot X^{(\alpha L)} \qquad (4.9) $$

in which $X^{(\beta L)}$ and $X^{(\alpha L)}$ denote the positions of an O-lattice point relative to the α and β crystallographic lattices and where ${}^{\alpha\beta}T$ is a linear transformation relating these two lattices. Evidently, the linear transformation matrix ${}^{\alpha\beta}T$ will be determined from the undeformed state of the solid, e.g. in terms of the orientation tensors ${}^{\alpha}O, {}^{\beta}O$. Hence the following relation (following Bollmann's work[36]) can be established:

$$(I - {}^{\alpha\beta}T^{-1}) \cdot X^{(0)} = b \tag{4.10}$$

where $X^{(0)}$ is the position of the O-point with respect to an orthogonal coordinate system and b the Burger vector pertaining to the αth lattice. The above relation is a generalization of the well-known Frank relation. It is, furthermore, an imaging relation between the b lattice of the αth crystal and the interpenetrating O-lattice. Although it may be analytically complicated, a unique solution for $X^{(0)}$ is possible. It is to be noted, however, that different O-elements are separated by "cell-walls" that usually bisect the line connecting O-points[102]. Hence, by constructing an O-lattice as well as the cell-walls between the O-elements, information of possible boundaries between the αth and βth crystal can be obtained. Once such a boundary is fixed, it will then represent arrays of atoms belonging to either the αth or βth crystal. The size and shape of the intersection between the boundary and cell-wall structure will then form the actual grain boundary coincidence cell as referred to in Chapter II.

Since a distinction has been made between low and high misfit orientations, it is to be noted that low misfit angle boundaries consist essentially of so-called "primary dislocations" which are positioned on the common line between any two grain boundary coincidence cells. However, between these dislocation lines there exist areas of crystalline material that become elastically distorted by the induced dislocation field, such that within the coincidence cell itself an O-point corresponds to a minimum strain. In the case of high misfit angle boundaries, where primary dislocations usually overlap, consideration of the dislocation effect becomes rather meaningless. However, one can construct in this case, on the assumption that some fraction of the structure in the sense of a coincidence lattice site is still retained, a "secondary dislocation network" with a different

Burger vector in the boundary zone. Thus one obtains from the first coincidence lattice site a second O-lattice based on the difference between the actual misfit angle and the nearest coincidence site lattice misfit angle. These are then the O_2-lattices mentioned earlier. Since the latter are somewhat larger than those due to primary dislocations, the resulting strain within the boundary becomes smaller. A more detailed discussion on such situations is given in reference[102].

In the light of the above remarks, the assumptions made for the model of grain boundary effects in the present theory can be summarized as follows:

(a) The relative crystal orientation corresponds to the coincidence cell orientation, e.g. the boundary zone is represented in terms of a periodic structure.

(b) Thus the boundary itself is a periodic extension of an array of atoms that forms an identifiable unit.

(c) The number of atoms as well as their positions are considered to vary.

(d) A central force interaction potential such as the Morse or Lenard–Jones potential, and also other types, can be employed to extract a "generalized surface traction" acting within the boundary zone.

(e) In general a relaxation of the position of atoms at the boundary may occur, but nearer to the core of the crystal the atoms retain their specific lattice sites.

In order to account for the mismatch angle which is significant for the microstructure of polycrystalline solids, the earlier formulation given in the kinematics of deformation of such materials (see Chapter II, Fig. 13) is somewhat extended here. Thus, it is convenient to introduce a "surface coordinate system" as shown in Fig. 17(b) for the contiguous α, β crystals with their respective body frames ${}^{\alpha}\boldsymbol{Y}, {}^{\beta}\boldsymbol{Y}$ since the latter meet under a certain mismatch orientation. If ${}^{\alpha}\boldsymbol{n}$ denotes the normal to the αth crystal surface, then in terms of the surface coordinates one can write the following relations:

$$ {}^{\alpha}\zeta_a: \quad {}^{\alpha}\zeta_1 = {}^{\alpha}\boldsymbol{n} \times {}^{\alpha\beta}\boldsymbol{\lambda}, \quad {}^{\alpha}\zeta_2 = {}^{\alpha}\boldsymbol{n}, \quad {}^{\alpha}\zeta_3 = {}^{\alpha\beta}\boldsymbol{\lambda} \tag{4.11} $$

in which the rotation about a common axis is expressed by the eigenvector ${}^{\alpha\beta}\boldsymbol{\lambda}$ as indicated in Fig. 17(b). Using the above algorithm and the assumptions (a)–(e), it is possible to formulate the grain boundary effect in terms of an energy that generally is a potential of the form $\Psi\{|{}^{\alpha\beta}\hat{\boldsymbol{d}}|\}$ as outlined earlier in this text. However, the form of this potential function strongly depends on the coincidence cell size and even more significantly on the relative displacement vector ${}^{\alpha\beta}\hat{\boldsymbol{d}}$ for the particular material under consideration. Since the coincidence areas (Fig. 14, Chapter II) are random functions of the orientation between the crystals, it follows that, without any loss of generality, the interfacial potential will also be a random function of these parameters. However, the choice of a suitable potential shows according to experimental evidence little effect on the structure of the grain boundary for a given material. Hence one may choose for instance a "Morse function" type of potential so that:

$$\Psi\{|{}^{\alpha\beta}\hat{\boldsymbol{d}}|\} = \Psi_0\{1-\exp[-b|{}^{\alpha\beta}\hat{\boldsymbol{d}}|]\}^2 \tag{4.12}$$

and where

$${}^{\alpha\beta}\hat{\boldsymbol{d}} = |{}^{\alpha\beta}\hat{\boldsymbol{d}}|\boldsymbol{e} \tag{4.13}$$

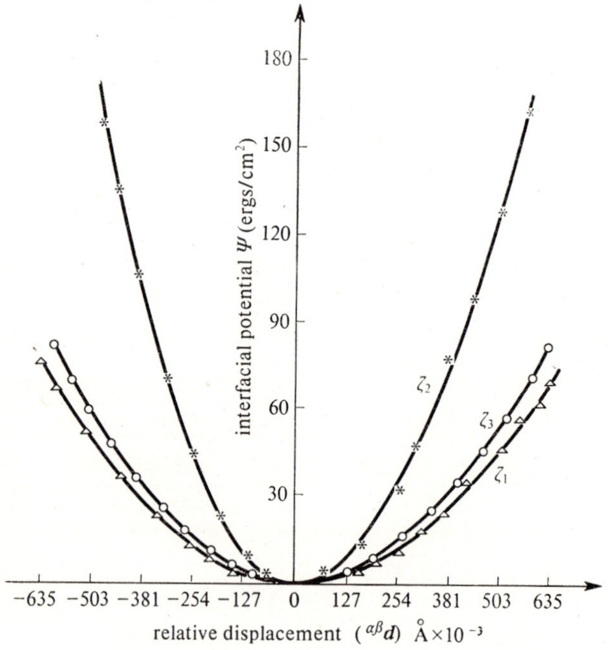

Fig. 18. Interfacial potential Ψ of copper (ergs/cm^2).

$\boldsymbol{e}$ being the unit vector in the direction of the relative displacement ${}^{\alpha\beta}\hat{\boldsymbol{d}}$ given earlier (equation (2.26), Chapter II). In this context the interfacial potential Ψ for a particular polycrystalline solid such as copper is given in Fig. 18 above. This interfacial potential is a surface potential derivable from the Morse function type of potential by a computer simulation technique[103]. It is plotted against the relative displacement ${}^{\alpha\beta}\hat{\boldsymbol{d}}$ (Å $\times 10^{-3}$) and illustrated by three curves corresponding to the reference coordinates of a coincidence cell point $(\zeta_1, \zeta_2, \zeta_3)$ as obtained from a computer simulation procedure for the grain boundary model discussed above (see also reference[103]). The two scalar quantities Ψ, Ψ_0 and the material characteristic constant b in relation (4.12) are obtainable from spectroscopic studies.

Following Yvon[38] a generalized force can be formulated by considering the discrete surface force between coincidence cell points (Fig. 14 (b)) such that:

$$ {}^{\alpha\beta}\hat{\boldsymbol{F}} = -\frac{\mathrm{d}\Psi\{|{}^{\alpha\beta}\hat{\boldsymbol{d}}|\}}{\mathrm{d}\{|{}^{\alpha\beta}\hat{\boldsymbol{d}}|\}}\boldsymbol{e} \tag{4.14} $$

which is visualized to act at each coincidence cell contained in the interface between the αth and βth crystal. Thus a generalized surface force accounting for the interaction between these crystals induced by the microstress field can be expressed in the sense of Gel'fand and Shilov[15] as follows:

$$ {}^{\alpha\beta}\boldsymbol{\tau}({}^{\alpha\beta}\boldsymbol{d}) = \langle \delta({}^{\alpha\beta}\boldsymbol{d} - {}^{\alpha\beta}\hat{\boldsymbol{d}}), {}^{\alpha\beta}\hat{\boldsymbol{F}} \rangle \tag{4.15} $$

in which $\delta(\cdot)$ is the three-dimensional Dirac-delta function and the brackets $\langle\ \rangle$ denote integration over an arbitrary volume containing the support of δ. Thus, ${}^{\alpha\beta}\boldsymbol{\tau}({}^{\alpha\beta}\boldsymbol{d}) = {}^{\alpha\beta}\boldsymbol{F}({}^{\alpha\beta}\hat{\boldsymbol{d}})|_{{}^{\alpha\beta}\hat{d} = {}^{\alpha\beta}d}$. It has been shown by Axelrad and Basu[82] that it is possible to arrive at an explicit form for the operator ${}^{\alpha\beta}\boldsymbol{B}(t)$ using the above generalized surface traction such that:

$$ {}^{\alpha}\boldsymbol{\mu} = {}^{\alpha\beta}\boldsymbol{B} : {}^{\alpha\beta}\boldsymbol{\tau}\,{}^{\alpha}\boldsymbol{n} \tag{4.16} $$

where ${}^{\alpha}\boldsymbol{\mu}$ is as stated previously the deformation in the interfacial surface. From the point of equilibrium between both crystals, it is evident that at this interface the following compatibility relations will hold, i.e.:

$$ {}^{\alpha\beta}\boldsymbol{\tau} = -{}^{\alpha\beta}\boldsymbol{\tau}, \quad {}^{\alpha}\boldsymbol{n} = -{}^{\beta}\boldsymbol{n} \tag{4.17} $$

Hence the interaction operator ${}^{\alpha\beta}\boldsymbol{B}$ will satisfy the following condition:

$$ {}^{\alpha\beta}\boldsymbol{B} = -{}^{\alpha\beta}\boldsymbol{B} \tag{4.18}$$

and by using the definition of the relative displacement ${}^{\alpha\beta}\boldsymbol{d}$ (equation (2.26)) it follows that:

$$ {}^{\alpha\beta}\boldsymbol{d} = -2\,{}^{\alpha\beta}\boldsymbol{B} : {}^{\alpha\beta}\boldsymbol{\tau}\,{}^{\alpha}\boldsymbol{n} \tag{4.19}$$

Combining (4.12), (4.14), (4.15) and (4.19) yields therefore an explicit form of the operator ${}^{\alpha\beta}\boldsymbol{B}$ as follows:

$$ {}^{\alpha\beta}\boldsymbol{B} = \frac{1}{2}\,{}^{\alpha\beta}\boldsymbol{d}\,{}^{\alpha}\boldsymbol{n}\left[\frac{2\Psi_0 b}{|{}^{\alpha\beta}\boldsymbol{d}|}\exp(-b|{}^{\alpha\beta}\boldsymbol{d}|) \times \{1-\exp(-b|{}^{\alpha\beta}\boldsymbol{d}|)\}\,{}^{\alpha\beta}\boldsymbol{d}\right] \tag{4.20}$$

In eqns. (4.16) and (4.19) the " : " sign designates a second order contraction.

(C) *Elastic response of a polycrystalline solid in tension*

In the preceding paragraphs explicit forms of the operators ${}^{\alpha}\boldsymbol{A}(t)$ and ${}^{\alpha\beta}\boldsymbol{B}(t)$ have been given (equations (4.8) and (4.20)). The significant parameters contained in the arguments of these operators can be determined for single microelements from experimental studies. However, it is extremely difficult to assess the relative displacement vector ${}^{\alpha\beta}\boldsymbol{d}$, although attempts are being made at present to find this quantity at least for the simpler case of fibrous materials from experimental techniques discussed in the subsequent chapter. It is suggested therefore in the case of polycrystalline solids to employ data based on the computer simulation method referred to previously. Hence in order to formulate the material operator for a microelement of crystalline solids a linear combination of ${}^{\alpha}\boldsymbol{A}(t)$ and ${}^{\alpha\beta}\boldsymbol{B}(t)$ or their inverse is used as an approximation in the following manner:

$$ \boldsymbol{M}(t) = [(1-k)\,{}^{\alpha}\boldsymbol{A}(t) + \Lambda k\,{}^{\alpha\beta}\boldsymbol{B}(t)] \tag{4.21}$$

or

$$ \boldsymbol{M}^{-1}(t) = [(1-k)\,{}^{\alpha}\boldsymbol{A}(t) + \Lambda k\,{}^{\alpha\beta}\boldsymbol{B}(t)]^{-1} \tag{4.22}$$

It is important to note that the above forms are only valid for specific structures and that in order to combine the internal effects with that due to the presence of a grain boundary in the formulation of the material operator $\boldsymbol{M}(t)$, a material characteristic quantity k has been introduced in (4.21) and (4.22). The parameter Λ in these expressions is a transformation matrix due to the transformation of the surface coordinates (4.11) to the external reference frame. The significance of the quantity k, however, should be further pointed out. Thus, for instance, if k is equated to zero in the above expressions, the operator $\boldsymbol{M}(t)$ becomes simply a transform operator describing the response of a single crystal or microelement. On the other hand, if k is equated to unity, the proposed material operator would express the effect of grain boundaries only. It is apparent that for an actual polycrystalline solid this parameter very likely will satisfy the condition $0 \leqslant k \leqslant 1$. In general, it will be a complicated function that depends on the lattice parameter of the crystal, the grain size as well as the mismatch angle between grain boundaries for a given microstructure. Although at present investigations are carried out to arrive at a rigorous functional form of k, it is beyond the scope of this text to give a more detailed analysis of it. To illustrate the general theory of stochastic deformations and its application to polycrystalline solids, two specific metals have been chosen, e.g. copper and aluminium for which arbitrary values of k lying between 0 and 1 will be used in the subsequent numerical analysis.

With regard to the transformation matrix Λ, it has been shown in reference[103], that by referring the local grain boundary coordinates ξ_a to the external fixed frame X_i that:

$$\xi_a = \alpha_{ia} X_i \tag{4.23}$$

so that a combination of α_{ia} coefficients in terms of the orientation angle φ (see Fig. 19) will give Λ whereby:

$$\alpha_{ia} \sim \begin{bmatrix} \cos\varphi & \sin\varphi & 0 \\ -\sin\varphi & \cos\varphi & 0 \\ 0 & 0 & 1 \end{bmatrix} \tag{4.24}$$

Fig. 19 also illustrates the polycrystalline solid subjected to a tensile load and several possible distributions of the angle φ. The simplest distribution will be obtained for the case when all crystals are oriented in one direction only. This case is of considerable interest in the analysis of high-temperature resistant materials and their application in engineering. Finally, in formulating the material operator $\boldsymbol{M}(t)$ in (4.21) and (4.22) it has been implicitly assumed that the microstress

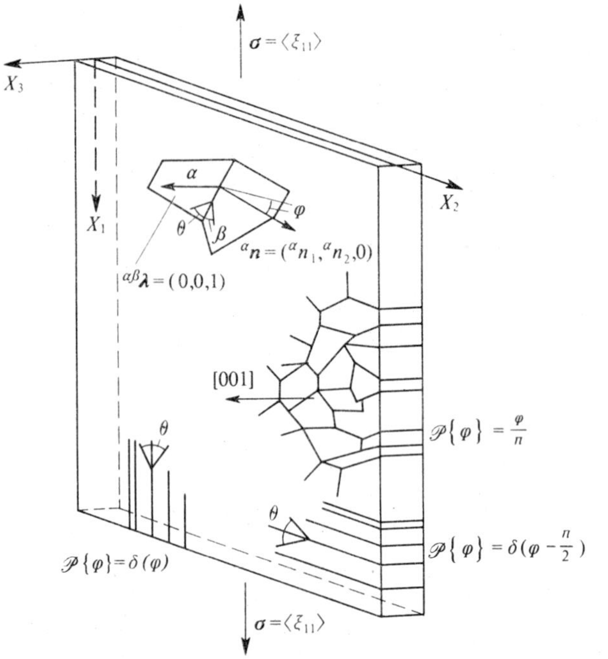

Fig. 19. Polycrystalline solid in tension.

across the boundary zone remains continuous, whereas the deformations are discontinuous. This assumption follows clearly from the considerations given in Section 2 and 3 of this chapter.

In order to obtain the required macroscopic response relations for the polycrystalline solid by using the concept of mesodomains in the material, it is necessary to find the statistical moments of the parameters contained in the material operator. Thus, for instance, considering first the operator ${}^{\alpha}\boldsymbol{A}(t)$ (equation 4.6), the first moment of that operator will be given by its mean value as follows:

$$\langle {}^{\alpha}\boldsymbol{A}(t) \rangle = \int {}^{\alpha}\boldsymbol{A}(t) \, \mathrm{d}\mathscr{P}\{\boldsymbol{\Gamma}\} \tag{4.25}$$

in which the only independent random variable is considered to be the orientation tensor $\boldsymbol{\Gamma}$. However, considering an alternative form of this operator in which a constant distribution of orientation has been admitted, and further assuming that the material characteristics such as G, ν and η remain the same for all crystals in the material, one can then write:

$$\langle {}^{\alpha}\boldsymbol{A}(t)\rangle = {}^{\alpha}\boldsymbol{A}(t) \tag{4.26}$$

It is significant for the operator ${}^{\alpha\beta}\boldsymbol{B}(t)$ to introduce the distribution of the mismatch orientation angle $\mathscr{P}\{\theta\}$ (see also Fig. 19), so that the first moment of the operator becomes:

$$\langle {}^{\alpha\beta}\boldsymbol{B}(t)\rangle = \int {}^{\alpha\beta}\boldsymbol{B}(t)\mathrm{d}\mathscr{P}\{\theta\} \tag{4.27}$$

It is further important to establish the distribution of the transformation tensor Λ in terms of $\mathscr{P}\{\boldsymbol{n}\}$ (see also reference[101]), whereby:

$$\langle \Lambda\rangle = \int \Lambda \mathrm{d}\mathscr{P}\{\boldsymbol{n}\} \tag{4.28}$$

and also the distribution of the material characteristic parameter k. However, the latter, for reasons given earlier, is at present not available and hence only arbitrarily assigned values of the same between $0 \leqslant k \leqslant 1$ can be given in the analysis. It is now possible to formulate an approximate expression for the material operator valid for a mesodomain and hence to construct the macroscopic response relations. Identifying the operator $\mathscr{M}(t)$ with its tensor representation we have:

$$\mathscr{M}(t) = \langle \boldsymbol{M}(t)\rangle = \langle [(1-k)^{\alpha}\boldsymbol{A}(t) + \Lambda k^{\alpha\beta}\boldsymbol{B}(t)]\rangle \tag{4.29}$$

On the assumption that the above quantities, i.e. ${}^{\alpha}\boldsymbol{A}$, ${}^{\alpha\beta}\boldsymbol{B}$, Λ and k are statistically independent and will assume constant values for a given microstructure, (4.29) can be written in a reduced form as follows:

$$\mathscr{M}(t) = (1-k)\langle {}^{\alpha}\boldsymbol{A}(t)\rangle + k\langle \Lambda\rangle \langle {}^{\alpha\beta}\boldsymbol{B}(t)\rangle \tag{4.30}$$

Following the above simplified forms, the response relations of the structured solid can be derived by noting that the distribution of the microstresses can be obtained from that of the microdeformations or strains by means of the distribution of the material operator $\boldsymbol{M}(t)$, i.e. $\mathscr{P}\{\boldsymbol{M}(t)\}$ (see equations (3.123) or (3.124)). It is evident that the latter distribution can be derived from that of the other parameters such as $\mathscr{P}\{\boldsymbol{\Gamma}\}$, $\mathscr{P}\{\theta\}$, $\mathscr{P}\{\boldsymbol{n}\}$, etc. Again to illustrate the case on hand, one can further simplify the form of the operator such that:

$$\mathscr{M}(t) = (1-k)^{\alpha}\boldsymbol{A}(t) + k^{\alpha\beta}\boldsymbol{B}(t) \tag{4.31}$$

whereby this simplification by no means limits the theory presented in this text, but is only used to point out the computational scheme one may follow to evaluate numerically the response behaviour of a polycrystalline solid. This will be further discussed below. Using the reduced form of the material operator above (4.31), the probability distribution of microstresses and deformations are then related by:

$$\mathscr{P}\{\xi\} = \mathscr{M}^{-1}(t)\mathscr{P}\{\boldsymbol{u}\} \tag{4.32}$$

so that the average microstress becomes:

$$\boldsymbol{\sigma} = \langle\xi\rangle = \mathscr{M}^{-1}(t)\langle\boldsymbol{u}\rangle \tag{4.33}$$

and the corresponding variance:

$$V(\xi) = \mathscr{M}^{-1}V(\boldsymbol{u})\mathscr{M}^{-1T} \tag{4.34}$$

(D) *Model analysis*

The material model used in the present analysis is illustrated by Fig. 19. As mentioned previously, only one particular distribution, i.e. $\mathscr{P}\{\varphi\} = \delta(\varphi)$ will be considered here, although several other distributions were used in reference[101]. To simplify the analysis and due to the availability of experimental data a two-dimensional model will be treated for the numerical analysis given below. Further, the normal microstress component ${}^{\alpha\beta}\xi$ in the grain boundary zone, which has been included in the investigation given in reference[101], is here deleted as well as the microdeformation component u_3. The latter is for all practical purposes taken as $u_3 = 0$. With these simplifications the operational equations concerning the response of a single crystal can be written as follows:

$$\begin{aligned} \xi_{11} &= M_{111}^{-1}u_1 + M_{112}^{-1}u_2 \\ \xi_{12} &= M_{121}^{-1}u_1 + M_{122}^{-1}u_2 \\ \xi_{22} &= M_{221}^{-1}u_1 + M_{222}^{-1}u_2 \end{aligned} \tag{4.35}$$

In these relations u_1 and u_2 are the components of the total deformation ${}^{\alpha}\boldsymbol{u}$ which includes as discussed earlier the internal deformation ${}^{\alpha}\boldsymbol{w}$ and the relative displacement vector ${}^{\alpha\beta}\boldsymbol{d}$. However, for the purpose of illustration only and since at present actual measurements of the components of ${}^{\alpha\beta}\boldsymbol{d}$ are not available, it is assumed for simplification that ${}^{\alpha}u_i \approx {}^{\alpha}w_i$. This does not mean, however, that the grain boundary effect has been neglected, but is considered only in terms of the material operator. Hence relations (4.35) can be expressed by the approximation:

$$\begin{aligned} \xi_{11} &= M^{-1}_{111} w_1 + M^{-1}_{112} w_2 \\ \xi_{12} &= M^{-1}_{121} w_1 + M^{-1}_{122} w_2 \\ \xi_{22} &= M^{-1}_{221} w_1 + M^{-1}_{222} w_2 \end{aligned} \tag{4.36}$$

Considering now the experimental data obtained for a two-dimensional model for AL-crystals (see reference[49]), the following expressions for the two components of the microdeformations can be written:

$$\begin{aligned} w_1 &= \frac{0.105 + 0.187\,|\sin\beta|}{132.96\,|\cos\beta|} \times 10^3(-\sin\beta + \cos\beta) + \Omega_1 \\ w_2 &= \frac{0.105 + 0.187\,|\sin\beta|}{132.96\,|\cos\beta|} \times 10^3(-\cos\beta + \sin\beta) + \Omega_2 \end{aligned} \tag{4.37}$$

in which these components are obtained in microns and where Ω_1, Ω_2 represent the deformation components of the centre of mass of the crystal as established by holographic interferometry, whilst β designates the angle of rotation of "Laue spots" due to the induced deformation and measured by means of an X-ray diffraction technique. These measurements will be more fully discussed in the following chapter. Hence, employing these techniques the distribution of ${}^{\alpha}\boldsymbol{w}$ or, equivalently, ${}^{\alpha}\boldsymbol{u}$ can be obtained and the stress distribution $\mathscr{P}\{\xi\}$ may be asserted by using relations (4.31) and (4.32). It should be noted, however, that in computing the material operator $\mathscr{M}$ or its inverse form (4.31), it has been implicitly assumed in the present simplified analysis that $\langle \Lambda \rangle = 1$. This conforms with the special distribution of $\mathscr{P}\{\varphi\} = \delta(\varphi)$, i.e. that the grain orientation is in the direction of the tensile loading, only. Continuing with this simplified model and using data for aluminium from references[94,100], the following material constants are adopted:

$$\begin{aligned} G &= 2.86 \times 10^{11} \text{ dynes/cm}^2 \\ \nu &= 0.347 \\ \eta &= 3.57 \times 10^{8} \text{ dynes/cm}^2 \end{aligned} \tag{4.38}$$

Using these material constants, the operator ${}^{\alpha}\boldsymbol{A}$ or its inverse ${}^{\alpha}\boldsymbol{A}^{-1}$ can be numerically determined. It is further convenient for computational purposes to rewrite the form (4.21) of the microelement operator as follows:

$$\boldsymbol{M}\colon M_{ijk} = (1-k)A_{ijk} + kB_{ijk} \tag{4.39}$$

where, in general,

$$A_{ijk} = \left[\frac{G+4\eta}{2G^2}\delta_{ik}\,\delta_{jl} - \frac{3G\nu+4\eta(1+\nu)}{6G^2(1+\nu)}\delta_{ij}\,\delta_{kl}\right]\nabla_l^{-1} \tag{4.40}$$

To compute B_{ijk} or its inverse B_{ijk}^{-1} in (4.39), equation (4.20) can be used with the understanding that in the absence of any experimental data of ${}^{\alpha\beta}\boldsymbol{d}$, the latter is taken on the basis of a computer simulation model of the grain boundary as mentioned previously. Thus Fig. 20 gives the values of ${}^{\alpha\beta}\xi$ plotted against ${}^{\alpha\beta}\boldsymbol{d}$ as obtained from the computer simulation method in terms of the three components ${}^{\alpha\beta}\zeta_1$, ${}^{\alpha\beta}\zeta_2$, ${}^{\alpha\beta}\zeta_3$. The slope of the lines corresponds in each case to the component of the third order tensor ${}^{\alpha\beta}\boldsymbol{B}$. However, as mentioned before to facilitate the computational scheme in the present formulation the contribution to the grain boundary effect is only considered in terms of the shear stress component ${}^{\alpha\beta}\xi_{12}$, whilst in the more general case (see also reference[101]) the normal component ${}^{\alpha\beta}\xi_{11}$ has also been taken into account. This, however, does not by any means restrict the general theory treated before. On the basis of the above remarks and denoting the bracketed expression in equation (4.40) by a fourth order tensorial operator A_{ijkl} such that:

$$A_{ijkl} = \left[\frac{G+4\eta}{2G^2}\delta_{ik}\,\delta_{jl} - \frac{3G\nu+4\eta(1+\nu)}{6G^2(1+\nu)}\delta_{ij}\,\delta_{kl}\right]. \tag{4.41}$$

The components of the material operator M_{ijk} can be written as follows:

$$\begin{aligned}
M_{111} &= (A_{1111}\nabla_1 + A_{1112}\nabla_2)\,(1-k) + kB_{111}\\
M_{112} &= (A_{1121}\nabla_1 + A_{1122}\nabla_2)\,(1-k) + kB_{112}\\
M_{121} &= (A_{1211}\nabla_1 + A_{1212}\nabla_2)\,(1-k) + kB_{121}\\
M_{122} &= (A_{1221}\nabla_1 + A_{1222}\nabla_2)\,(1-k) + kB_{122}\\
M_{221} &= (A_{2211}\nabla_1 + A_{2212}\nabla_2)\,(1-k) + kB_{221}\\
M_{222} &= (A_{2221}\nabla_1 + A_{2222}\nabla_2)\,(1-k) + kB_{222}
\end{aligned} \tag{4.42}$$

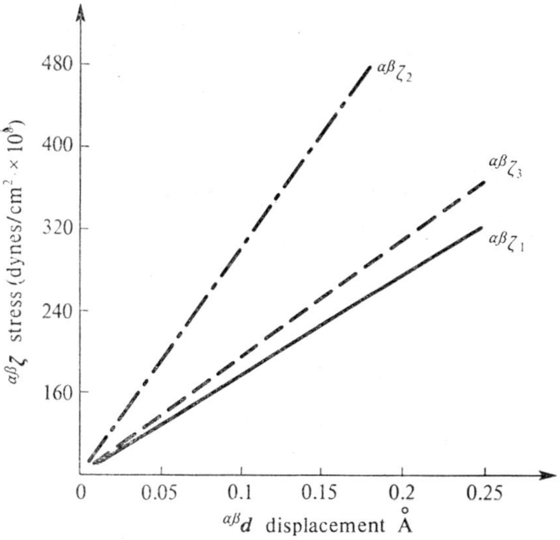

Fig. 20. Grain boundary stress ${}^{\alpha\beta}\xi$ against relative displacement ${}^{\alpha\beta}\boldsymbol{d}$ (F.C.C.-AL crystal, 10^8 dynes/cm^2).

Hence, using the material constants (4.38) and the components of ${}^{\alpha\beta}\boldsymbol{B}$ in accordance with Fig. 20, the numerical values obtained from a computer analysis of (4.42) for the components of the material operator $\boldsymbol{M}$ are given in Table IV below by assigning the material characteristic parameter k values between $0 \leqslant k \leqslant 1$ as discussed earlier.

TABLE IV

Component of the microelement operator $\boldsymbol{M}$ (F.C.C.-AL Crystal, 10^{13} dynes/cm^2)

Material characteristic k	0	0.1	0.3	1.0
M_{111}	0.60937	0.54844	0.42656	0.0
M_{112}	0.32479	0.29231	0.22736	0.0
M_{121}	0.0	0.97720×10^{-11}	0.29316×10^{-10}	0.9772×10^{-10}
M_{122}	0.28458	0.25612	0.19921	0.2238×10^{-9}
M_{221}	0.32479	0.29231	0.22736	0.0
M_{222}	0.60937	0.54844	0.42656	0.0

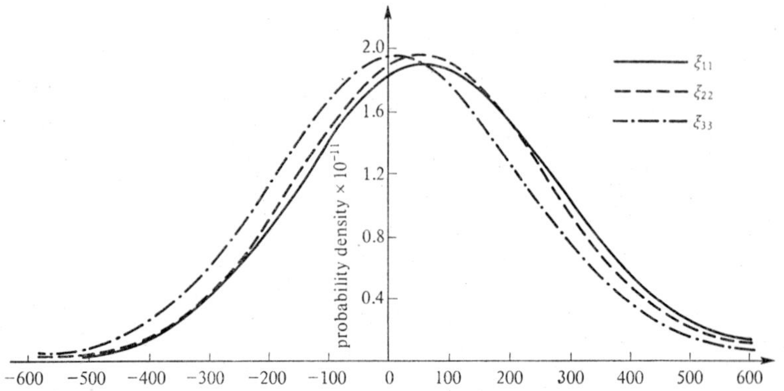

Fig. 21. Microstress distribution ($k = 0$); AL-crystal, 10^8 dynes/cm^2.

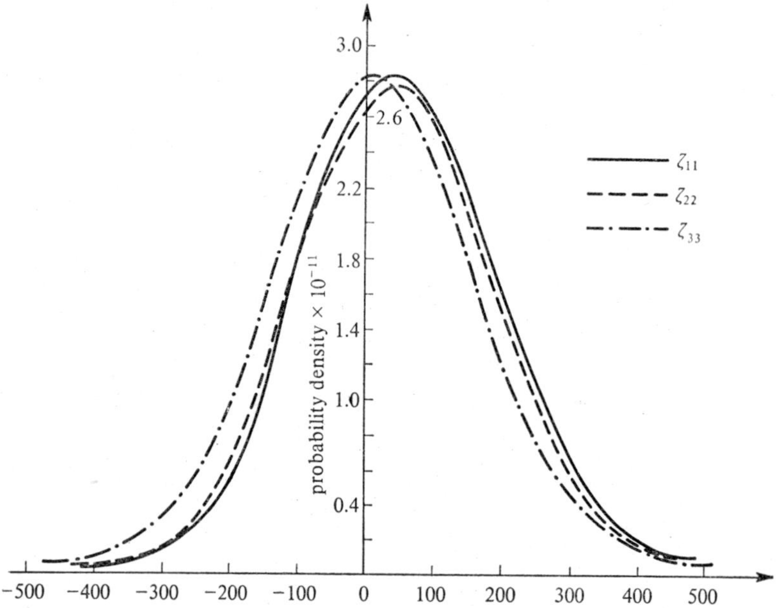

Fig. 22. Microstress distribution with grain boundary effect ($k = 0.3$) AL-crystal 10^8 dynes/cm^2.

Following the interpretation of the significance of k given previously, the microstress distribution for the case $k = 0$, i.e. without the effect of grain boundaries is shown in Fig. 21, where the curves correspond to the three components $\xi_{11}, \xi_{12}, \xi_{22}$ (equations (4.35) or (4.36)). Another plot of the microstress involving the effect of grain boundaries to the extent as dealt with in the model used in the present analysis is given in Fig. 22 for the case of $k = 0.3$.

4.3 Mechanical response of fibrous systems

Another application of the general theory of stochastic deformations concerning fibrous systems will be considered in this section. Although it may appear that such systems and polycrystalline solids are basically different, there are nevertheless, similarities. Thus in both cases microelements can be conveniently defined and interaction effects can be accounted for within the framework of the theory. Whilst some remarks with regard to such systems have been made in Chapter I (Figs. 4, 6) and Chapter II (Fig. 16), it is necessary for the subsequent analysis to discuss in some detail the "single fibre" behaviour and the "bonding" effects between them. It will be shown that a material operator for a structural element can be established from which by using its distribution within a specific mesodomain, macroscopic response relations can be derived. For the illustration of the theory and in order to present a numerical evaluation of the required response relations a two-dimensional fibrous network will be considered. This simplification from the more general three-dimensional case is mainly due to the limited availability of experimental information with regard to single fibres and the bond behaviour.

(A) *Single fibre behaviour*

It has been stated in Chapters I and II that a microelement of a fibrous system is considered to consist of a certain "fibre segment" located between two half bonding areas between neighbouring fibres. In order to arrive at a material operator for such an element first the single fibre and then the bond behaviour must be investigated. With reference to the deformation kinematics of an element as given in Chapter II (Figs. 15, 16), the deformation occurring in a fibre segment (equation (2.22)) is rewritten here, so that:

$$^f u_I : {}^f\boldsymbol{u}(t) = {}^f\boldsymbol{y}(t) - {}^f\boldsymbol{Y} \tag{4.43}$$

in which the superscript "*f*" as indicated previously refers to the "fibre segment" and the subscript "*I*" to the internal coordinate frame (Y_1, Y_2, Y_3), so that:

$$^f u_i = {}^f\Lambda_{Ii}\, {}^f u_I \tag{4.44}$$

where Λ_{Ii} is the transformation matrix from the internal to the external fixed reference frame (X_1, X_2, X_3). Introducing the base vectors of the body frame $^\alpha \boldsymbol{g}_I$ and that for the external frame $\boldsymbol{e}_i$ permits to write an approximate strain measure for the elastic deformation of the fibre segment as follows:

$$^f\boldsymbol{e} : {}^f e_{IJ} = \frac{1}{2}(\delta_{Ik}\nabla_J + \delta_{Jk}\nabla_I)\, u_k \tag{4.45}$$

The above relation is however a simplification of a possible strain measure so far as real fibrous materials are concerned, since other microstructural effects will make the single fibre itself inhomogeneous and bring about a more complex measure. In the present theory the fibre segment is treated as a homogeneous medium and only the most significant characteristic, i.e. its rheological behaviour is taken into account. To characterize this behaviour of a single fibre one can use the phenomenological description in terms of hereditary integrals (see for instance reference[104]) as follows:

$$^f\boldsymbol{e}(t) = {}^f\boldsymbol{E}^{-1}\left[{}^f\boldsymbol{\xi}(t) + \int_{-\infty}^{t} {}^f\boldsymbol{H}(t-\tau)\, {}^f\boldsymbol{\xi}(\tau)\, \mathrm{d}\tau\right] \tag{4.46}$$

$$^f\boldsymbol{\xi}(t) = {}^f\boldsymbol{E}\left[{}^f\boldsymbol{e}(t) - \int_{-\infty}^{t} {}^f\boldsymbol{R}(t-\tau)\, {}^f\boldsymbol{e}(\tau)\, \mathrm{d}\tau\right] \tag{4.47}$$

in which $^f\boldsymbol{E}$ denotes the elastic tensor modulus of the fibre, $^f\boldsymbol{H}$, $^f\boldsymbol{R}$ the tensorial creep and relaxation functions and the time-dependent microstress and microstrain in the fibre segment, respectively. It is important to note that experimental observations show that almost all fibre materials exhibit a non-linear rheological behaviour. Hence the above response relations (4.46), (4.47) in the case of uni-axial loading corresponding to the available test data can be rewritten as follows:

$$
{}^{f}e(t) = g_1[{}^{f}\xi(t), l_1, l_2, \ldots]+ + \int_{0^+}^{t} h_1({}^{f}\xi(t), c_1, c_2, \ldots) H(t-\tau)\mathrm{d}\tau \tag{4.48}
$$

$$
{}^{f}\xi(t) = g_2[{}^{f}e(t), k_1, k_2, \ldots]+ + \int_{0^+}^{t} h_2({}^{f}e(t), b_1, b_2, \ldots) R(t-\tau)\mathrm{d}\tau \tag{4.49}
$$

in which the functions $g_1(\cdot)$ and $g_2(\cdot)$ correspond to the non-linear elastic response, whilst the functions $h_1(\cdot)$ and $h_2(\cdot)$ represent the non-linear hereditary effects. It is seen that the arguments of these functions involve unknown constants designated by $l_1, l_2, \ldots, k_1, k_2, \ldots, c_1, c_2, \ldots$ and $b_1, b_2, \ldots$, which can be determined by means of a differential approximation method discussed below (see also Distefano and Todeschini[105]). Thus, considering for example the functions $g_1({}^{f}\xi(t), l_1, l_2, \ldots)$ and $g_2({}^{f}e(t), k_1, k_2, \ldots)$ to be linear ones and that for the case of stress relaxation where ${}^{f}e(t) = {}^{f}e(0^+) = e_i$ during the ith experiment, one can rewrite relation (4.49), since $h_2(\ldots)$ is independent of time, as follows:

$$
{}^{f}\hat{\xi}_i(t) = {}^{f}Ee_i + h_2(e_i, b_1, b_2, \ldots) \int_{0^+}^{t} R(\tau)\mathrm{d}\tau \tag{4.50}
$$

However, to find an expression for the relaxation kernel $R(\tau)$ above, it is assumed that the latter will satisfy an Nth order differential equation as follows:

$$
a_0 R + a_1 R^{(1)} + a_2 R^{(2)} + \ldots + a_{N-1} R^{(N-1)} + a_N R^{(N)} = 0 \tag{4.51}
$$

in which

$$
R^{(I)} = \frac{\mathrm{d}^I R}{\mathrm{d}t^I}, \quad (I = 1, \ldots, N) \tag{4.52}
$$

It can be shown that the general solution of (4.51) is given by:

$$
R(t) = \sum_{I=1}^{N} D_I \exp(F_I t) \tag{4.53}
$$

where F_I $(I = 1, \ldots, N)$ are the N roots of (4.51) and the D_I's are constants. Hence equation (4.50) can also be written as follows:

$$^f\xi_i = {}^fEe_i + h_2(e_i, b_1, b_2, \ldots) \sum_{I=1}^{N} \frac{D_I}{F_I} \{\exp(F_I t) - 1\} \tag{4.54}$$

Defining the second term on the right-hand side of this relation as an operator $R^*(e_i, t)$ such that:

$$R^*(e_i, t) = -h_2(e_i, b_1, b_2, \ldots) \sum_{I=1}^{N} \frac{D_I}{F_I} \{\exp(F_I t) - 1\} \tag{4.55}$$

permits to write the stress-strain relation in operational form as:

$$^f\xi_i(t) = A(t)\,{}^fe_i \tag{4.56}$$

where

$$A(t) = {}^fE - R^*(e_i, t) \tag{4.57}$$

It is possible to generalize relation (4.56) to hold for any experiment and all components of $^f\xi(t)$ and $^fe(t)$ so that the general response relation becomes:

$$^f\xi(t) = {}^\alpha A(t) : {}^fe(t) \tag{4.58}$$

in which

$$^\alpha A(t) : {}^\alpha A_{ijkl}(t) = {}^fE_{ijkl} - R^*{}_{ijkl}(e_{ij}, t) \tag{4.59}$$

The constitutive relation (4.58) can also be expressed in terms of a relation between microstress and microdeformation such that:

$$^f\xi(t) = {}^fA(t) \cdot {}^f\boldsymbol{u}(t) \tag{4.60}$$

Thus by using (4.58) and (4.59) the stress-deformation relation can be expressed by:

$$^f\xi(t) = {}^fA(t) \cdot {}^f\boldsymbol{u}(t) = [{}^fE - {}^f\boldsymbol{R}^*(t)] : \nabla \boldsymbol{u}(t) \tag{4.61}$$

The above expression will be employed later in formulating the material operator for a microelement of the fibrous structure. The gradient operation as mentioned previously is taken with respect to the attached body frame.

Returning to the uni-axial formulation, in order to evaluate the unknown constants $b_1, b_2, \ldots$ in the function $h_2({}^fe_i(t), b_1, b_2, \ldots)$ it is now necessary to use an optimization procedure. For this purpose consider the following analytical expression:

$$\bar{\xi}_i(t) = Ee_i + G_i \theta_i(t, m_{i1}, m_{i2}, \ldots) \qquad (4.62)$$

in which $\bar{\xi}_i(t)$ is an "experimental stress" as a function of time, G_i are experimental material constants and $\theta_i(t, m_{i1}, m_{i2}, \ldots)$ corresponds to the form of the integral of the relaxation kernel Γ_i that can be obtained from experimental data. It is convenient to define a functional $f(b_1, b_2, \ldots, D_I/F_I)$ containing the unknown constants so that:

$$f(b_1, b_2, \ldots, D_I/F_I) = \sum_{i=1}^{N} \gamma_i \int_{0^+}^{t} ({}^f\hat{\xi}_i(t) - {}^f\bar{\xi}_i(t))^2 \mathrm{d}t \qquad (4.63)$$

in which γ_i is a weighting function in the ith experiment. The optimization problem now involves the minimization of (4.62) subject to the constraint that:

$$\sum_{i=1}^{N} \int_{0^+}^{t} \{(a_0 \Gamma_i + a_1 \Gamma_i^{(1)} + \ldots) - (a_0 R + a_1 R^{(1)} + \ldots)\}^2 \mathrm{d}t \qquad (4.64)$$

is a minimum for all possible choices of the constants $a_0, a_1, \ldots, a_N$. A detailed discussion of this procedure is given in reference[106].

(B) *Bond behaviour*

In order to formulate the response behaviour of a fibrous system, it is important to include the bonding interaction. Previous studies concerned with the bond behaviour between fibres have been carried out mainly from a phenomenological point of view. Such considerations were based on measurements of the ultimate bond strength in the junction area of two overlapping fibres. In the present theory such experimental observations and their results are of little consequence and other experimental techniques are at present being developed to achieve measurements related to the significant parameter ${}^{\alpha\beta}\boldsymbol{d}$ involved in the bond deformation. These techniques will be briefly discussed in the following chapter. For the interpretation of measurements of the latter type a brief discussion on bond behaviour for the case of "perfect bonding", e.g. for a strongly idealized situation is given below. Due to the fact that one has to distinguish between an "actual bonding area" and one which corresponds to an optically observed "total bonding area", the relation between these areas will be:

$$^{B}A = \eta\,^{T}A \tag{4.65}$$

in which the factor η represents the number of bonds that may be available in the case of perfect bonding such that:

$$\eta = \frac{^{B}A}{^{T}A} = \frac{^{B}N}{2\,^{T}N} \tag{4.66}$$

where ^{B}N is the number of total bonds in the area ^{T}A and the factor 2 accounts for the two matching sides of the bonded area. Evidently the ratio η is only obtainable from experimental observations of a specific fibrous structure. It is convenient for the subsequent considerations to use the "area per bond" defined by:

$$^{b}a = \frac{^{T}A}{^{B}N} \tag{4.67}$$

as well as the "unit cell area a" as indicated for the idealized model of a hydrogen bonding between two fibres α, β in Fig. 23. With reference to the above figure one can consider two matching points 2–2′ that belong to the "hydroxyl group" within the unit cell and which are separated in the undeformed state of the material by the previously used distance vector $^{\alpha\beta}\boldsymbol{\Delta}$ (Fig. 23 (a)). The latter as already discussed in Chapter II, is in this theory considered as the most significant parameter so far as the deformation kinematics are concerned. In the deformed configuration of the bond this vector changes to $^{\alpha\beta}\boldsymbol{\delta}$ corresponding to a relative displacement between points 2 to 2″ (Fig. 23 (b)) so that with reference to the attached body frame:

$$^{b}d_I(t): {}^{b}\boldsymbol{d}(t) = {}^{b}\boldsymbol{\delta}(t) - {}^{b}\boldsymbol{\Delta} \tag{4.68}$$

and by application of the transformation matrix Λ_{Ii}:

$$^{b}d_i(t): {}^{b}\boldsymbol{d}(t) = {}^{\alpha}\Lambda_{Ii}\,{}^{b}d_I(t) \tag{4.69}$$

In view of the importance of the bonding potential and its effect on the bonding in the junction area between the fibres, it is necessary to consider this type of bonding in more detail. A more rigorous study of hydrogen bonding can be found in references[107,108], where

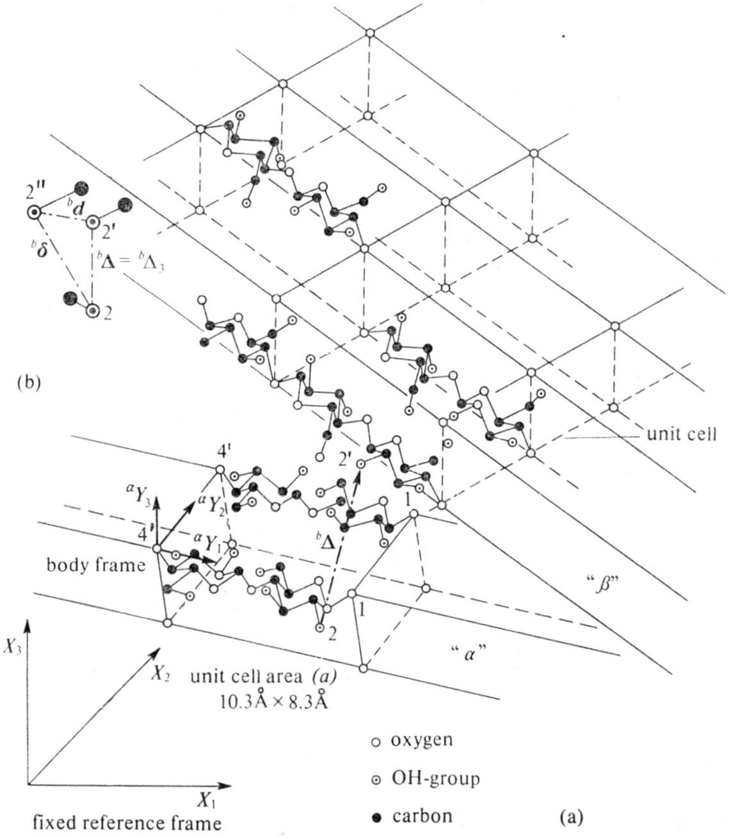

Fig. 23. Idealized bonding of a fibre-fibre interface (three-dimensional).

also a classification of this type of bonding is given. In the present application of the theory this type of bonding occurring in natural fibres can be illustrated by a bond potential plotted against the relative deformation as shown in Fig. 24 (a, b).

It may be seen from this figure that the presence of a free oxygen atom nearest to the OH-group in the αth fibre itself causes a slight shift of the potential representing the actual covalent bonding between the O- and H-atom (curves 1, 2 in Fig. 24 (b)). It is assumed in the present model that such a free oxygen atom in the αth fibre surface matches with a corresponding O-atom in the OH-group in the matching surface of the βth fibre. Hence, by neglecting the actual shift due to the presence of a free oxygen atom in the αth fibre, the potential assumed to exist is of the type indicated in Fig. 24 (c). On the basis of this assumption and the idealized model of the fibre-fibre interface for a three-dimensional situation (Fig. 23), it is possible to introduce a "Morse type potential" that involves the relative displacement vector associated with an actual equilibrium position during the deformation of two bonded fibres such that:

$$^{b}\Psi(t) = {^{b}\Psi_0}\{\exp(-2\nu|^{b}\boldsymbol{d}(t)|)-2\exp(-\nu|^{b}\boldsymbol{d}(t)|)\} \qquad (4.70)$$

in which $^{b}\Psi_0$ is the equilibrium potential, ν a material characteristic and $^{b}\boldsymbol{d}(t)$ the relative displacement to be established from experi-

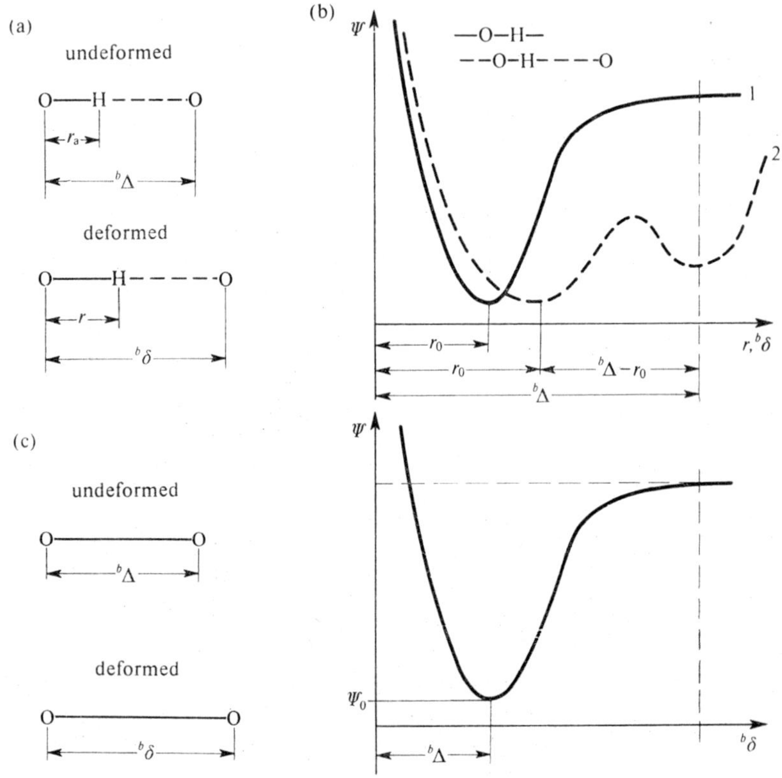

Fig. 24. Bonding potential for hydrogen bonding between natural fibres.

mental investigations discussed in the following chapter. It is important to note, however, that the above form of the potential[109] is rather incomplete from a quantum theoretical point of view[110], since it neglects a second term in the expression that is due to the repulsion energy between the hydrogen and free oxygen atoms as well as a third term caused by the exchange energy of attraction between the oxygen atoms (see, for instance, Pauling[107]). From a micromechanics point of view, the potential form (4.70) is considered acceptable as an approximation, so that a formulation of the mechanical response behaviour of fibrous structures can be achieved. As a consequence of this remark and that given earlier concerning the number of hydrogen bonds within a junction area, it is evident that each area per bond ${}^b a$ will be acted upon by a force vector ${}^b\boldsymbol{f}(t)$ associated with the corresponding microstress in the bond as follows:

$$ {}^b\xi(t) = {}^b\boldsymbol{n}\frac{{}^b\boldsymbol{f}(t)}{{}^b a} \tag{4.71} $$

where ${}^b\boldsymbol{n}$ is the outward unit normal to the bonding area between the αth and βth fibre. Employing this form and the earlier one giving the discrete interaction force (4.14) leads to the relation between ${}^b\boldsymbol{f}(t)$ and ${}^b\Psi(t)$ for the case of a fibrous structure as follows:

$$ \begin{aligned} {}^b\boldsymbol{f}(t) &= -\left[\frac{\partial\, {}^b\Psi(t)}{\partial\,|{}^b\boldsymbol{d}(t)|}\right]{}^b\boldsymbol{e} \\ &= 2\nu\, {}^b\Psi_0\{\exp(-2\nu\,|{}^b\boldsymbol{d}(t)|)-\exp(-\nu\,|{}^b\boldsymbol{d}(t)|)\}\,{}^b\boldsymbol{e} \end{aligned} \tag{4.72} $$

in which ${}^b\boldsymbol{e}$ is the unit base vector employed to define the force vector ${}^b\boldsymbol{f}(t)$ with reference to the local coordinate frame ${}^\alpha Y_I$. Thus the stress in the bonding area between two matching fibres can be written as:

$$ {}^b\xi(t) = \frac{2\nu\, {}^b\Psi_0}{{}^b a}\{\exp(-2\nu\,|{}^b\boldsymbol{d}(t)|)-\exp(-\nu\,|{}^b\boldsymbol{d}(t)|)\}\,{}^\alpha\boldsymbol{n}\,{}^b\boldsymbol{e} \tag{4.73} $$

It is seen that the interaction operator so far as this type of bonding is concerned is a function of the vector ${}^b\boldsymbol{d}(t)$ so that in operational form the interaction effect can be expressed implicitly by:

$$ {}^b\xi(t) = \mathscr{B}[{}^b|\boldsymbol{d}(t)|] \tag{4.74} $$

It is apparent that the above operator has an analogous meaning to that representing the interaction effect for grain boundaries of polycrystalline solids. However, in the present case the operator results from considerations of the bonding effect between two hydroxyl groups symbolized by the matching points 2–2′ in the bonding area of two overlapping fibres. It is to be noted that in order to obtain an explicit stress-deformation relation according to (4.74) in component form, it is necessary to consider the appropriate components of the stress ${}^{b}\xi(t)$ and the relative displacement ${}^{b}\boldsymbol{d}(t)$ as discussed below.

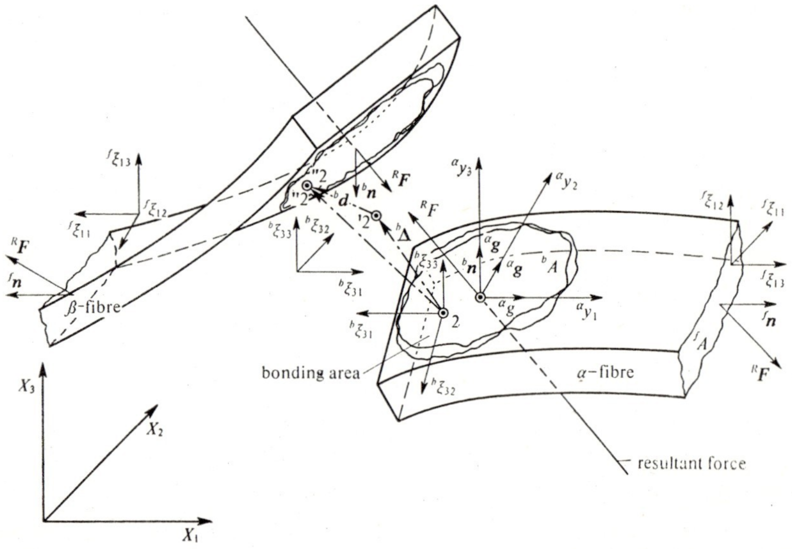

Fig. 25. Microstress components in the microelements and bonding area.

An illustration for the general case of bond response is indicated in Fig. 25. Assuming that the outward normal (equation 4.71) to the junction area coincides with the unit base vector ${}^{b}e_3$, i.e. that ${}^{b}\boldsymbol{n}$ has the components $(0, 0, 1)$ the corresponding microstress components will be ${}^{b}\xi_{31}(t)$, ${}^{b}\xi_{32}(t)$ and ${}^{b}\xi_{33}(t)$ as indicated in the figure. Thus considering one of these components, for example, ${}^{b}\xi_{31}(t)$

and using an asymptotic expansion of the exponential term in the expression (4.73) by following the exposition given in reference[106], the relative displacement component ${}^{b}d_1(t)$ corresponding to that stress, becomes:

$$
\begin{aligned}
{}^{b}\xi_{31}(t) = \frac{2\nu^{b}\Psi_0}{{}^{b}a}\Bigg\{\Bigg[1+(-2\nu^{b}d_1(t))+\frac{(-2\nu^{b}d_1(t))^2}{2!}+\ldots+ \\
+\frac{(-2\nu^{b}d_1(t))^{n-1}}{(n-1)!}\Bigg]-\Bigg[1+(-\nu^{b}d_1(t))+ \\
+\frac{(-\nu^{b}d_1(t))^2}{2!}+\ldots+\frac{(-\nu^{b}d_1(t))^{n-1}}{(n-1)!}\Bigg]\Bigg\}\cdot{}^{b}n_3{}^{\alpha}g_1
\end{aligned}
\tag{4.75}
$$

in which each of these series in the bracket converges to the exponential function given in (4.73) in the interval $-\infty < |{}^{b}\boldsymbol{d}(t)| < \infty$. It is possible to use for simplification the first two terms of each series only and consequently (4.75) reduces then to:

$$
{}^{b}\xi_{31}(t) = \frac{-2\nu^{2b}\Psi_0}{{}^{b}a}\cdot{}^{b}n_3{}^{\alpha}g_1{}^{b}d_1(t) \tag{4.76}
$$

The above relation represents one-component of the more general and explicit form of the stress-deformation relation (4.74), such that:

$$
{}^{b}\boldsymbol{\xi}(t) = {}^{b}\boldsymbol{B}(t)\cdot{}^{b}\boldsymbol{d}(t). \tag{4.77}
$$

in which the component ${}^{b}B_{311}$ of the operator ${}^{b}\boldsymbol{B}(t)$ becomes:

$$
{}^{b}B_{311} = -2\frac{\nu^{2b}\Psi_0}{{}^{b}a} \tag{4.78}
$$

It is important to note that in the above model of bonding (Fig. 25) it has been implicitly assumed that whilst the components of the bond stress may vary from point to point within the bond area, the resultant or integrated values of the former can be replaced by a resultant force vector ${}^{R}\boldsymbol{F}$ and a moment ${}^{R}\boldsymbol{M}$. This would then correspond to a fibre-fibre bonding situation which exhibits under uniaxial loading bending and shear effects. However, due to the inaccessibility of more fundamental experimental data of such effects, it is only possible at present to adopt the much simpler model indicated by relations (4.72)–(4.78) that will be discussed subsequently in the context of the numerical analysis of cellulosic structures. It is further significant to recognize that if the induced microstresses in the bond areas become large enough a disruption or breakage of the assumed "perfect bonding" between the fibres will occur. In order to describe analytically such a phenomenon from the point of view of probabilistic micromechanics it is evident that more information on the significant parameter ${}^{b}\boldsymbol{d}(t)$ must be obtained first. For this purpose experimental studies to obtain distribution functions of this random quantity are at present being carried out, which will be more fully discussed in the following chapter.

(C) *Response behaviour of a microelement of a fibrous structure*

In order to formulate the mechanical response of a structural element of the fibrous material (Fig. 15, Chapter II), it is necessary to consider the contribution to the microstress for both parts of the element, i.e. the fibre segment and the bonding area. In a similar manner as discussed for polycrystalline solids, a material characteristic function will have to be used in the present case as well. Since such a function is essentially a time-dependent one due to the fact that the actual bonding area will change with time and under the application of a stress, it is assumed in this analysis that the function can be expressed by:

$$ {}^{\alpha}K(t) = \frac{{}^{b}A(t)}{{}^{\alpha}L \times W} \tag{4.79} $$

where ${}^{\alpha}K(t)$ will be valid for a specified width of the fibre segment "W" and its length "${}^{\alpha}L$" as well as the actual bonding area ${}^{b}A(t)$. Hence analogously to the treatment of crystalline solids the total deformation ${}^{\alpha}\boldsymbol{u}(t)$ of a microelement becomes:

$$ {}^{\alpha}\boldsymbol{u}(t) = [1-{}^{\alpha}K(t)]^{\alpha}\boldsymbol{A}\cdot{}^{f}\boldsymbol{u}(t)+{}^{\alpha}K(t)^{\alpha}\boldsymbol{A}\cdot{}^{b}\boldsymbol{d}(t) \tag{4.80} $$

in which the orientation matrix ${}^{\alpha}\boldsymbol{A}$ transforms the local components of ${}^{f}\boldsymbol{u}(t)$ and ${}^{b}\boldsymbol{d}(t)$ to the external reference frame X_I. It is seen that for the case of fibrous structures, the function ${}^{\alpha}K(t)$ is somewhat simpler than in the case of polycrystalline solid in that it is of a purely geometrical nature. Thus using the relation for the microstress in the fibre segment (4.61) and the relation for ${}^{b}\boldsymbol{d}(t)$ (4.77) one can write the total deformation of a microelement as follows:

$$ \begin{aligned} {}^{\alpha}\boldsymbol{u}(t) = {} & [1-{}^{\alpha}K(t)]^{\alpha}\boldsymbol{A}\cdot{}^{f}\boldsymbol{A}^{-1}(t):{}^{f}\xi(t)+ \\ & +{}^{\alpha}K(t)^{\alpha}\boldsymbol{A}\cdot{}^{b}\boldsymbol{B}^{-1}(t):{}^{b}\xi(t) \end{aligned} \tag{4.81} $$

However, to express the overall response relation in operational form, it becomes necessary to establish a relation between the microstress ${}^{f}\xi(t)$ pertaining to the fibre segment and ${}^{b}\xi(t)$ acting in the bond area. Since the stress components in the bonding area have been considered in the above model (Fig. 25) to be equipollent to the resultant force ${}^{R}\boldsymbol{F}$, the components of the latter by neglecting the moment effects can be written as:

$$\begin{aligned} {}^{R}F_1 &= \sum_{i=1}^{{}^{b}N} ({}^{b}f_1)_i = \sum_{i=1}^{{}^{b}N} ({}^{b}\xi_{31}{}^{b}a)_i = \iint_{{}^{f}A} {}^{f}\xi_{11}\,\mathrm{d}^{\alpha}y_2\,\mathrm{d}^{\alpha}y_3 \\ {}^{R}F_2 &= \sum_{i=1}^{{}^{b}N} ({}^{b}f_2)_i = \sum_{i=1}^{{}^{b}N} ({}^{b}\xi_{32}{}^{b}a)_i = \iint_{{}^{f}A} {}^{f}\xi_{12}\,\mathrm{d}^{\alpha}y_1\,\mathrm{d}^{\alpha}y_3 \\ {}^{R}F_3 &= \sum_{i=1}^{{}^{b}N} ({}^{b}f_3)_i = \sum_{i=1}^{{}^{b}N} ({}^{b}\xi_{33}{}^{b}a)_i = \iint_{{}^{f}A} {}^{f}\xi_{13}\,\mathrm{d}^{\alpha}y_1\,\mathrm{d}^{\alpha}y_2 \end{aligned} \tag{4.82}$$

Thus the relation between microstresses in the fibre segment and that in the bonding area can be approximated by:

$$\sum_{i=1}^{{}^{b}N} ({}^{b}\xi^{b}a)_i = \iint_{{}^{f}A} {}^{f}\xi\,\mathrm{d}^{\alpha}y_2\,\mathrm{d}^{\alpha}y_3 \tag{4.83}$$

With this simplifying assumption and recognizing that the cross-sectional area of the fibre segment is ${}^{f}A$, one gets:

$${}^{b}\xi(t)^{b}A = {}^{f}\xi(t)^{f}A \tag{4.84}$$

However, as has been pointed out previously the actual bonding area is also a function of time, the above relation in its simplest form can be rewritten so that:

$${}^{b}\xi(t) = \frac{{}^{f}A}{{}^{b}A(t)}\,{}^{f}\xi(t) = {}^{\alpha}\varkappa(t)^{f}\xi(t) \tag{4.85}$$

in which the new parameter ${}^{\alpha}\varkappa(t)$ is assumed to be accessible from experimental studies and on the assumption that it will be independent of the stress in the fibre segment. On the basis of the above relation and the underlying assumptions it is now possible to express the material operator for a structural element as follows:

$${}^{\alpha}\boldsymbol{M}(t) = [(1-{}^{\alpha}K(t))^{\alpha}\boldsymbol{A}\cdot{}^{f}\boldsymbol{A}^{-1}(t)+{}^{\alpha}K(t)^{\alpha}\boldsymbol{A}\cdot{}^{b}\boldsymbol{B}^{-1}(t)^{\alpha}\varkappa(t)] \tag{4.86}$$

or in operational form:

$$\begin{aligned} {}^{\alpha}\boldsymbol{u}(t) &= {}^{\alpha}\boldsymbol{M}(t) : {}^{\alpha}\xi(t) \\ {}^{\alpha}\xi(t) &= {}^{\alpha}\boldsymbol{M}^{-1}(t) \cdot {}^{\alpha}\boldsymbol{u}(t) \end{aligned} \tag{4.87}$$

(D) *Response behaviour of a fibrous system*

In order to obtain the macroscopic relation for a fibrous system, it is necessary to consider the distribution of the material operator ${}^{\alpha}\boldsymbol{M}$, as well as that of the microstresses and deformations. In complete analogy to crystalline solids one can write:

$$\mathscr{P}\{{}^{\alpha}\boldsymbol{u}(t)\} = \mathscr{P}\{{}^{\alpha}\boldsymbol{M}(t)\}\mathscr{P}\{{}^{\alpha}\xi(t)\} \tag{4.88}$$

in which the distribution functions can be expressed to a first approximation in terms of the first and second moments of the relative quantities, valid over a specific mesodomain of the material sample. In the case of fibrous materials the choice of a specific mesodomain as to its size depends largely on the available experimental technique which makes the establishment of the distribution functions possible. A more detailed discussion on such a choice of the mesodomain is given below in the "model analysis" for the fibrous structure. Thus the macroscopic values or first moment or the microdeformations for instance, can be obtained as follows:

$${}^{M}\boldsymbol{u}(t) = {}^{M}\langle \boldsymbol{u}(t)\rangle = \sum {}^{\alpha}\boldsymbol{u}(t)\Delta\mathscr{P}\{{}^{\alpha}\boldsymbol{u}(t)\} \tag{4.89}$$

Similarly the mean value of the microstresses giving the macroscopic stress over the mesodomain can be expressed by:

$${}^{M}\boldsymbol{\sigma}(t) = {}^{M}\langle \xi(t)\rangle = \sum {}^{\alpha}\xi(t)\Delta\mathscr{P}\{{}^{\alpha}\xi(t)\} \tag{4.90}$$

Noting that the mesodomain-material operator can be obtained in a similar manner as shown in Section 4.2, the latter will be:

$${}^{M}\mathscr{M}(t) = {}^{M}\langle {}^{\alpha}\boldsymbol{M}(t)\rangle = \sum {}^{\alpha}\boldsymbol{M}(t)\Delta\mathscr{P}\{{}^{\alpha}\boldsymbol{M}(t)\} \tag{4.91}$$

so that the macroscopic response relation assumes the usual form presented earlier as follows:

$${}^{M}\boldsymbol{u}(t) = {}^{M}\mathscr{M}(t) : {}^{M}\boldsymbol{\sigma}(t) \tag{4.92}$$

Since a complete description of these quantities must also contain the second moments, the variances can be expressed as follows:

$$\begin{aligned} V(\boldsymbol{u}(t)) &= {}^{\mathrm{M}}\mathcal{M}(t)\, V(\xi(t)^{\mathrm{M}}\mathcal{M}^{\mathrm{T}}(t) \\ V(\xi(t)) &= {}^{\mathrm{M}}\mathcal{M}^{-1}(t)\, V(\boldsymbol{u}(t))^{\mathrm{M}}\mathcal{M}^{-1}(t)^{\mathrm{T}} \end{aligned} \tag{4.93}$$

where "T" indicates the transpose and the inverse operator ${}^{M}\mathcal{M}^{-1}(t)$ has been used, leading to the stress-deformation relation:

$${}^{M}\sigma(t) = {}^{M}\mathcal{M}^{-1}(t) \cdot {}^{M}\boldsymbol{u}(t) \tag{4.94}$$

The use of this operator will be further shown in the "model analysis" below. It is, however, assumed in the above formulation that the material operator is statistically independent from the other relevant quantities characteristic of the microstructure of the system.

(E) *Model analysis*

The mechanical response relations for a three-dimensional fibrous network has been briefly discussed in the foregoing paragraphs. However, due to the experimental data available at present for two-dimensional systems only, the application of the stochastic deformation theory will be shown for such a system, when it is subjected to a tensile loading. Thus the material model for a "cellulosic network" is indicated in Fig. 26 showing the network to be subjected to a uni-axial tension in the X_1-direction and a mesodomain bounded by two "theoretical scanning lines" $s_1 - s_1$ and $s_2 - s_2$. It is seen that the mesodomain chosen as indicated in the figure consists of a number of "scanning areas" of dimensions $\mu \times \mu$, where for cellulosic systems μ is frequently of the order of 1–2 mm. This subdivision of the meso-domain into individual "scanning areas A" is necessary so that experimental observations of the same under an electron microscope can be linked to observations by means of holographic interferometry in which a theoretical scanning line consists of a number of "points" rather than distinct areas. To clarify this further consider a particular area A of the mesodomains as shown in Fig. 27. It is seen that the centre of mass of such an area (C.M.) differs from the geometrical centre of the latter (G.C.) by a vector ${}^{\alpha}\boldsymbol{J}$.

Thus by considering the position vectors of the geometrical centres

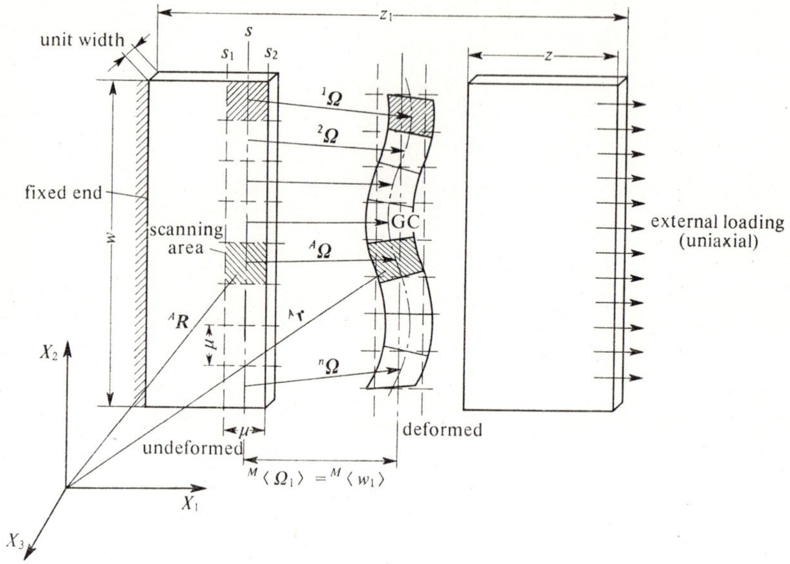

Fig. 26. Mesodomains of a two-dimensional model of a fibrous network under uni-axial loading.

with respect to the fixed coordinate frame in the undeformed and deformed configurations of the scanning area, respectively, it is seen that:

$$^{A}\boldsymbol{\Omega}(t) = {}^{A}\boldsymbol{r}(t) - {}^{A}\boldsymbol{R} \tag{4.95}$$

where the vector $^{A}\boldsymbol{\Omega}(t)$ of the geometric centre of A is measurable by means of holographic interferometry in much the same way as discussed previously for polycrystalline solids. The techniques used for such measurements will be more fully described in the following chapter. It may be noticed from Fig. 27 that for the deformed configuration the centre of mass differs from the geometrical centre of A by a vector $^{\alpha}\boldsymbol{j}(t)$ and in the undeformed state by a vector $^{\alpha}\boldsymbol{J}$. Hence the vector $^{\alpha}\boldsymbol{\Omega}(t)$ can be expressed by:

$$^{\alpha}\boldsymbol{\Omega}(t) = {}^{A}\boldsymbol{\Omega}(t) + {}^{\alpha}\boldsymbol{j}(t) - {}^{\alpha}\boldsymbol{J} \tag{4.96}$$

where the difference vector $^{\alpha}\boldsymbol{j}(t) - {}^{\alpha}\boldsymbol{J}$ is accessible from two successive observations of the area A under the scanning electron microscope and hence $^{\alpha}\boldsymbol{\Omega}(t)$ becomes measurable. Recalling from relation (2.27) (Section 2.4, Chapter II) that the total deformation of a microele-

Undeformed

Deformed

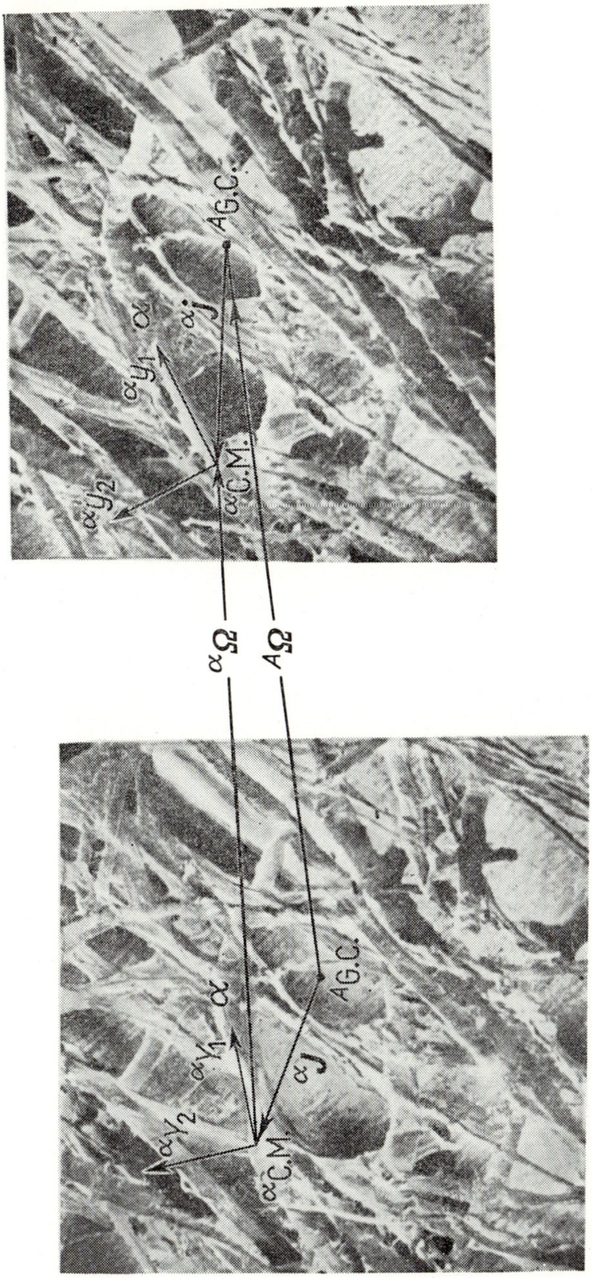

Fig. 27. Micrograph with magnification of ×100 for a scanning area *A* on the surface of a tensile specimen of a two-dimensional fibrous sheet in the undeformed and deformed states under the scanning electron microscope.

ment can be expressed with reference either to the external frame (${}^{\alpha}\boldsymbol{w}(t)$) or with respect to the attached body frame (${}^{f}\boldsymbol{u}(t)$), one can establish a relation between these quantities and the above vector ${}^{\alpha}\boldsymbol{\Omega}(t)$ on the assumption that nearly "perfect bonding" will prevail for the case of small deformations:

$$ {}^{\alpha}\boldsymbol{w}(t) = {}^{f}\boldsymbol{u}(t) + {}^{\alpha}\boldsymbol{\Omega}(t) \tag{4.97} $$

It is implicitly assumed in this relation that an overall deformation of the structural element occurs which is essentially caused by the action of the fibres and that the likelihood of bond deformations is small. This, however, is not the case in actual fibrous materials, but can be looked upon as a first approximation for the treatment in the present model analysis. It follows therefore that one can write:

$$ {}^{A}\langle {}^{\alpha}\boldsymbol{w}(t)\rangle = {}^{A}\langle {}^{f}\boldsymbol{u}(t)\rangle + {}^{A}\langle {}^{\alpha}\boldsymbol{\Omega}(t)\rangle \tag{4.98} $$

Furthermore, in order to illustrate the overall response behaviour of a two-dimensional fibrous network and in view of the restriction imposed by the present holographic interferometry technique in which the actual scanning areas can only be considered as points, it can be stated that:

$$ {}^{A}\langle {}^{f}\boldsymbol{w}(t)\rangle \approx {}^{A}\boldsymbol{\Omega}(t) \tag{4.99} $$

This statement amounts to an approximation required to overcome the "scale factor" which is naturally involved by combining the holographic interferometry technique with that of the scanning electron microscopy. A more careful assessment of such a scale factor would lead to a more rigorous relation than that given by (4.99). For the present purpose, however, by using this approximation and expression (4.98) a relation between the deformations of fibre segments and the earlier mentioned "difference vector" can be established as follows:

$$ {}^{A}\langle {}^{f}\boldsymbol{u}(t)\rangle = {}^{A}\langle {}^{\alpha}\boldsymbol{J} - {}^{\alpha}\boldsymbol{j}(t)\rangle \tag{4.100} $$

For the subsequent numerical analysis of a fibrous system such as cellulosic structure for which only limited experimental data are available, it is important to acquire a more specific expression for the significant stochastic parameter ${}^{\alpha}K(t)$ mentioned earlier in (4.79). In the latter relation this parameter has been shown to be time dependent since the actual bonding area is also a function of time. It has also been remarked that if the induced microstresses become large enough a "bond breakage" will occur. In the present model analysis however, as a matter of simplification of the analysis, it is assumed that in spite of the rheological characteristic of the single fibre, bond breakage has not developed to such an extent that it must be included in the overall response behaviour of the network. In this sense the essential contribution to the mechanical response is taken to be due to the fibre response itself and that the bonding parameter ${}^{\alpha}K(t)$ as a function of ${}^{b}A(t)$ can be considered as time-independent. This restriction on the model used cannot be removed before more quantitative experimental data become available. In this context, a new

TABLE V

Material parameters of a structural element of natural cellulose

Single fibre parameters	
Fibre segment length ${}^{\alpha}L$	
mean value	94.0×10^{-4} cm
standard deviation	36.111×10^{-4} cm
Fibre width W	37.5×10^{-4} cm
Fibre thickness h	1.5×10^{-4} cm
Elastic modulus E	450.0×10^{8} dyne/cm^2
Relaxation parameter b (for $N = 4$);	
function $h_2 = \text{const.}$	80.534×10^{-2}
Relaxation coefficient D_I/F_I $(I = 4)$	34.748×10^{8} dyne/cm^2
Orientation ${}^{\alpha}\Phi$: $({}^{\alpha}\mathbf{g}_1, \mathbf{e}_1)$	
mean value	-0.1134 radian
standard deviation	1.1257 radian
Bonding parameters	
Bonding ratio η	1.0
Actual bonded area ${}^{b}A$	
mean value	2000×10^{-8} cm^2
standard deviation	1147×10^{-8} cm^2
Relative bonded area ${}^{\alpha}K$	
mean value	0.5270
standard deviation	0.1194
Parameter ${}^{\alpha}\varkappa$ (ratio between bond stress and fibre stress)	0.0510
Equilibrium potential ${}^{b}\Psi_0$	3.14×10^{5} dyne Å
Spectroscopic constant ν	2 Å^{-1}
Unit cell area a	10.35 Å $\times$ 8.35 Å

technique for obtaining such data concerning the bond behaviour between two natural fibres will be briefly discussed in Section 5.5 of the following chapter.

In accordance with the above simplifying assumptions, the material parameters pertaining to a structural element of a cellulose structure are given in Table V. It may be observed that so far as the single fibre parameters are concerned, mean values of their lengths, widths and thicknesses have been taken as well as a simplifying relation to that given by (4.55), whereby for the relaxation parameter the function h_2 $(e_i, b_1, b_2, \ldots)$ has been considered as constant. Furthermore, the relaxation coefficient D_I/F_I has been evaluated by using the general relation for the relaxation kernel for the number of roots $N = 4$ $(I = 4)$. The orientations of the single fibres ${}^{\alpha}\Phi$ with respect to the X_1-direction, i.e. in the direction of the uni-axial loading of the material sample has been evaluated from a so-called "polar diagram". Table V also contains in accordance with these simplifications the average values of the actual bonded area, relative bonded area ${}^{\alpha}K$ and the parameter ${}^{\alpha}\varkappa$, etc., where these quantities correspond to the assumption of a nearly "perfect bonding", e.g. that the contribution of the bond response is negligible in the present model analysis.

In order to achieve a relation between the geometrical properties of the fibrous structure and the fundamental physical quantity of a sheet of such material known as the "basis weight" a more specific expression for the stochastic parameter ${}^{\alpha}K$ must be established. Conventionally the basis weight, which is directly related to the mass distribution in a fibrous network is obtained for the case of cellulosic networks from measurements by means of microdensitometers and β-radiography. Considering in this context the micrographs of Fig. 27, it is possible to define the basis weight for the cellulose network as follows:

$$
{}^{A}\Xi = \frac{{}^{A}m}{\mu^2} = \frac{{}^{A}U\,{}^{A}\lambda\varrho}{\mu^2} \tag{4.101}
$$

in which ${}^{A}\Xi$ is the "basis weight", ${}^{A}m$ the total weight of the structure within the scanning area A, ${}^{A}U$ the number of "crossings" of fibres in that area, ${}^{A}\lambda$ an average length of the cellulose fibres and ϱ the weight per unit length of an individual fibre. It should be noted that the average length of fibres is the expected value obtained from the probability distribution of "interception" between any two fibres lying in the scanning area. In this context a "statistical geometry of fibrous networks" has been discussed in reference[111], where the interception or fibre crossings for straight line fibre elements has been expressed as follows:

$$^{A}N_c = \tfrac{1}{2}\,{}^{A}P\,{}^{A}U[{}^{A}U-1] \tag{4.102}$$

in which ^{A}P denotes the probability of intersection between two fibres or fibre segments $^{\alpha}\lambda, {}^{\beta}\lambda$ within the scanning area and which include angles $^{\alpha}\Phi$, $^{\beta}\Phi$, respectively, with a fixed direction such as X_1, for example (see Fig. 27). The above relation has also been given in a somewhat similar form for curved fibres or fibre segments by Andreichenko[112]. Hence in terms of these orientations the probability distribution can also be expressed by:

$$^{A}P = \frac{({}^{A}\lambda)^2}{\mu^2}\sin|{}^{\alpha}\Phi-{}^{\beta}\Phi|, \quad \langle{}^{\alpha}\lambda\rangle = {}^{A}\lambda \tag{4.103}$$

It is to be noted that the random nature of the fibrous structure can be accounted for in a more general manner by introducing a distribution function $\Lambda({}^{\alpha}\lambda)$ of the individual fibre lengths as well as an angular distribution function $\mathscr{P}\{{}^{\alpha}\Phi\}$ of the angles $^{\alpha}\Phi$. However, such a general formulation leads to a more complicated functional relation that becomes difficult to assess experimentally. It is more convenient to employ a "scanning method" by crossing the scanning area in several different directions and then obtain the number of intersections within the area in terms of a periodic function or Fourier series. In this manner, by retaining the first two terms of such a series, one can express the angular distribution as follows:

$$\mathscr{P}\{{}^{\alpha}\Phi\} = \frac{1}{\pi}+e\cos(2{}^{\alpha}\Phi) \tag{4.104}$$

in which the double angle of $^{\alpha}\Phi$ has been introduced for convenience and where the first term on the right-hand side of (4.104) corresponds to the value of the distribution function for a completely random or non-oriented structure for which the distribution is always circular. The second term in this relation contains a constant "e" known as the eccentricity from the radial distribution and is obtained from measurement of actual "machine produced" cellulose structures. Since the scanning method is usually carried out in various directions radially from an arbitrarily chosen origin in the plane of the material sample, the eccentricity e depends on the two coefficients of the truncated Fourier series, as well as on the negative angle which the chosen direction, i.e. X_1 for example, makes with the origin. Using the method of least squares for the determination of the constants yields then a value e. This value varies between zero for a completely random network and $-1/\pi$ for a highly oriented one. The distribution curves obtained in this manner are known, as mentioned previously, as "polar diagrams"[113].

Thus, using the random orientation of a fibre within the scanning area and the above definition of the eccentricity e, leads via the distribution function (4.104) to an average of interception per fibre as follows:

$$^{A}n_c = \frac{(^{A}\lambda)^2}{\mu^2}\,{}^{A}U\left[\frac{1}{\pi} - e^2\frac{\pi}{6}\right] \tag{4.105}$$

Using the above relation and equation (4.101) permits to establish a relation between the length of a fibre segment $^{\alpha}L$ and the basis weight of the structure as:

$$^{\alpha}L = \frac{\varrho}{^{A}\Xi\left(\dfrac{1}{\pi} - e^2\dfrac{\pi}{6}\right)} \tag{4.106}$$

Following the argument given previously for the function $^{\alpha}K(t)$ to be constant for small deformations and nearly perfect bonding, yields an approximate relation between this parameter and the basis weight as follows:

$$^{\alpha}K(t) = \frac{^{b}A}{W}\left[\frac{^{A}\Xi}{\varrho}\left(\frac{1}{\pi} - e^2\frac{\pi}{6}\right)\right] \tag{4.107}$$

It is now possible, on the basis of this simplified formulation and on the additional assumption that the ratio between the cross-sectional area of the fibre and bonding area remains constant in the case of small deformations, to express the material operator for a structural element by:

$$
{}^{\alpha}\boldsymbol{M}(t) = \left[\left\{1 - \frac{{}^{b}A^{A}\Xi}{\varrho}\left(\frac{1}{\pi} - e^{2}\frac{\pi}{6}\right)\right\}{}^{\alpha}\boldsymbol{\Lambda}\cdot{}^{f}\boldsymbol{A}^{-1}(t) + \right.
$$

$$
\left. + \frac{{}^{f}A^{A}\Xi}{W\varrho}\left(\frac{1}{\pi} - e^{2}\frac{\pi}{6}\right){}^{\alpha}\boldsymbol{\Lambda}\cdot{}^{b}\boldsymbol{B}^{-1}\right] \tag{4.108}
$$

Using the above form of the material operator and recognizing that for a two-dimensional fibrous system in uni-axial loading, the micro-stresses and deformations will be given by:

$$
{}^{\alpha}\xi \sim \begin{bmatrix} {}^{\alpha}\xi_{11} & 0 & 0 \\ 0 & 0 & 0 \\ 0 & 0 & 0 \end{bmatrix}, \quad {}^{\alpha}\boldsymbol{u} \sim ({}^{\alpha}u_1, {}^{\alpha}u_2, 0) \tag{4.109}
$$

permits to express the response relation of a structural element given earlier in a more general form (4.87), for the present case as follows:

$$
\begin{aligned}
{}^{\alpha}u_1 &= {}^{\alpha}M_{111}{}^{\alpha}\xi_{11} & {}^{\alpha}\xi_{11} &= {}^{\alpha}M_{111}^{-1}{}^{\alpha}u_1 \\
{}^{\alpha}u_2 &= {}^{\alpha}M_{211}{}^{\alpha}\xi_{11} & {}^{\alpha}\xi_{11} &= {}^{\alpha}M_{211}^{-1}{}^{\alpha}u_2
\end{aligned} \tag{4.110}
$$

in which the operators ${}^{\alpha}M_{111}$ and ${}^{\alpha}M_{211}$ can be written more explicitly from (4.108) as follows:

$$
\begin{aligned}
{}^{\alpha}M_{111} &= \left[\left\{1 - \frac{{}^{b}A^{A}\Xi}{\varrho}\left(\frac{1}{\pi} - e^{2}\frac{\pi}{6}\right)\right\}{}^{\alpha}\Lambda_{11}{}^{f}A_{111}^{-1}(t) + \right. \\
&\left. + \frac{{}^{f}A^{A}\Xi}{W\varrho}\left(\frac{1}{\pi} - e^{2}\frac{\pi}{6}\right){}^{\alpha}\Lambda_{11}{}^{b}B_{311}^{-1}\right] \\
{}^{\alpha}M_{211} &= \left[\left\{1 - \frac{{}^{b}A^{A}\Xi}{\varrho}\left(\frac{1}{\pi} - e^{2}\frac{\pi}{6}\right)\right\}{}^{\alpha}\Lambda_{12}{}^{f}A_{111}^{-1}(t) + \right. \\
&\left. + \frac{{}^{f}A^{A}\Xi}{W\varrho}\left(\frac{1}{\pi} - e^{2}\frac{\pi}{6}\right){}^{\alpha}\Lambda_{12}{}^{b}B_{311}^{-1}\right]
\end{aligned} \tag{4.111}
$$

A more detailed discussion on the derivation of the above relations (4.111) is given in reference[106]. By using the previously given data (Table V) and the forms of the operators ${}^{f}A_{111}^{-1}(t)$, ${}^{b}B_{311}^{-1}$ a more explicit expression of the latter can be given as follows:

$$
{}^{f}A_{111}(t) = \left[{}^{f}E - b\sum_{I=1}^{N}\frac{D_I}{F_I}\{\exp(F_I t) - 1\}\right] \tag{4.112}
$$

$$
{}^{b}B_{311} = \frac{-2\nu^{2}\,{}^{b}\Psi_{0}}{{}^{b}a}\,{}^{b}n_3{}^{\alpha}g_1\cdot{}^{\alpha}g_1^{-1} \tag{4.113}
$$

The input data required for the numerical analysis of the operators (4.111), (4.112) and (4.113) have been taken from holographic interferometry tests on cellulose structures under tensile loading as described in Section 5.4 of the following chapter. In modelling the bonding behaviour on the basis of a two-dimensional material sample only one relative displacement component of the vector $^b\boldsymbol{d}$, i.e. $^bd_1$ has been considered. This corresponds to a bond motion of the matching point 2′ to 2″ in Fig. 23 and can only be regarded as an approximation, which serves to illustrate the present model analysis. It is further necessary to calculate the elements $^\alpha\Lambda_{11}$ and $^\alpha\Lambda_{12}$ of the transformation matrix $^\alpha\Lambda$ involved in expressions (4.111), which can be obtained from:

$$^\alpha\Lambda_{Ii} \sim \begin{bmatrix} \cos(^\alpha g_1, e_1) & \cos(^\alpha g_1, e_2) & 0 \\ \cos(^\alpha g_2, e_1) & \cos(^\alpha g_2, e_2) & 0 \\ 0 & 0 & 1 \end{bmatrix} \tag{4.114}$$

Table VI below contains the numerical values of the operators $^fA_{111}(t)$, $^bB_{311}(t)$, $^\alpha M_{111}(t)$ and $^\alpha M_{211}(t)$ for the purely elastic response as well as for the rheological response due to the fibres only, since the bond behaviour has been considered to be nearly perfect

TABLE VI

Material operators and their inverses for the single fibre, the bond and the microelement of a fibrous network

Time Sec.	$^fA_{111}$ 10^{13} $\frac{\text{dyne}}{\text{cm}^3}$	$^fA^{-1}_{111}$ 10^{-13} $\frac{\text{cm}^3}{\text{dyne}}$	$^bB_{311}$ 10^{20} $\frac{\text{dyne}}{\text{cm}^3}$	$^bB^{-1}_{311}$ 10^{-20} $\frac{\text{cm}^3}{\text{dyne}}$	M_{111} 10^{-13} $\frac{\text{cm}^3}{\text{dyne}}$	M^{-1}_{111} 10^{13} $\frac{\text{dyne}}{\text{cm}^3}$	M_{211} 10^{-13} $\frac{\text{cm}^3}{\text{dyne}}$	M^{-1}_{211} 10^{13} $\frac{\text{dyne}}{\text{cm}^3}$
0	0.76	1.31	0.35	2.85	0.61	1.62	0.08	12.06
36	0.74	1.35	0.35	2.85	0.66	1.53	0.09	11.17
72	0.71	1.41	0.35	2.85	0.69	1.45	0.09	10.23
108	0.68	1.47	0.35	2.85	0.72	1.40	0.10	9.94
144	0.66	1.52	0.35	2.85	0.75	1.35	0.10	9.52
180	0.64	1.56	—	—	0.77	1.39	0.11	9.31
216	0.61	1.64	—	—	0.78	1.31	0.11	9.14
252	0.59	1.70	—	—	0.80	1.30	0.11	8.92
288	0.57	1.75	—	—	0.81	1.29	0.11	8.83

for the cases of small deformations. In this context the small time increment shown in the table corresponds to that required during the holographic interferometry tests, where tensile testing of structures of this type must be performed before the onset of a noticeable creep behaviour (see also Section 5.4, Chapter V).

Within the framework of the simplifying assumptions made in this analysis and the given geometrical and physical properties of elements of the cellulose structure, the probability density of microstresses within a mesodomain is indicated in Fig. 28. Thus curve (a) in the figure shows the Gaussian character of the density function, which for small time increments and considerations of the viscoelastic behaviour of the fibres in the structure experiences a slight shift (curves (b), (c)). Such a shift will of course be more pronounced for a longer duration of time and will then represent the actual creep or relaxation behaviour of fibrous systems.

The analytical formulation of such a response behaviour has been

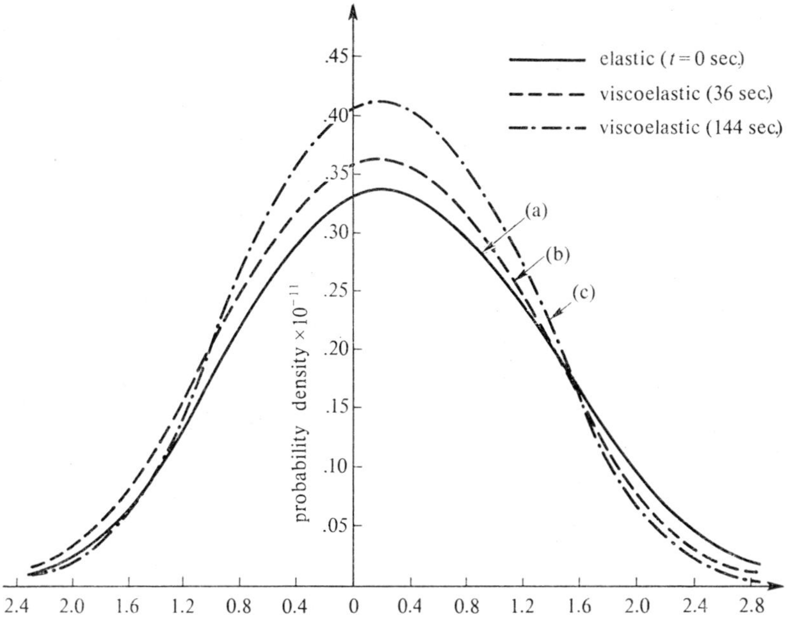

Fig. 28. Microelement stresses 10^{11} dyne/cm^2 in a cellulose system.

given earlier in references[114,115], whilst a more advanced study is at present being undertaken[116]. The corresponding experimental investigations will be briefly discussed in the following chapter. Finally, by using the above microstress distribution, the stress ${}^{M}\sigma(t)$ (equation 4.90) for a mesodomain as well as the material operator ${}^{M}\mathscr{M}(t)$ (4.91) can be evaluated and a stress-deformation relation for cellulosic structures can be established (4.94) corresponding to the conventional form of macroscopic response relations.

4.4 Mechanical relaxation of crystalline solids

As a final application of the general theory of stochastic deformations the mechanical relaxation of crystalline solids is discussed in this section. Amongst the many relaxation phenomena occurring in such materials the mechanical relaxation is an important one. In continuum mechanics this phenomenon is usually investigated to a first approximation by an after-effect theory[117] in which the mechanical response of the material is expressed in terms of the macroscopic quantities. In view of the theory developed in this text, which includes microstructural effects, stochastic models of relaxation phenomena are considered on the basis of probabilistic concepts[118]. Previous investigations concerned with crystalline solids and fibrous structures have considered such an approach in detail. It has been shown, in particular, for the case of crystalline solids[83] that one has to introduce in the formulation of the mechanical relaxation different types of relaxation functions, i.e. those valid for the interior of the microelement of the structure and those which represent the relaxation of the interaction zone between elements. It is intended in the present analysis to give a unified formulation of the internal and grain boundary relaxation of a polycrystalline solid (see also references[119,120]).

(A) *Rheological field quantities*

In accordance with the concepts of probabilistic micromechanics, the microstress acting on a single element of the crystalline solid can be regarded as a stochastic "stimulus" and the corresponding deformation as a stochastic "response". Hence, as already discussed earlier, it can be written that[82]:

$$\begin{aligned} {}^{\alpha}\xi &= {}^{\alpha}\xi({}^{\alpha}X;\, t) \\ {}^{\alpha}\boldsymbol{u} &= {}^{\alpha}\boldsymbol{u}({}^{\alpha}X;\, t) \end{aligned} \tag{4.115}$$

in which the symbols have the meaning given in Chapter II. It is assumed within an incremental theory and for the case of small deformations, that ${}^{\alpha}\boldsymbol{u}(t)$ is an additive function of the undeformed position vector X_k ($k = 1, 2, 3$) of the element as defined by relations (2.30) and (2.31). It has been mentioned in Chapter II that the mechanical relaxation behaviour of structured solids will be discussed more fully in this section and hence an analogous form for the "total deformation ${}^{\alpha}\boldsymbol{u}(t)$" (equation 2.32) is given below. It is convenient for the present analysis to introduce the superscripts "i" and "s", respectively, where the former will refer to the internal deformations and the latter to the deformation occurring in the grain boundary zone. Hence, using the additivity property relation (2.32) of Chapter II, can be rewritten to read:

$$\begin{aligned} {}^{\alpha}\boldsymbol{u}(t) = {} & [H(|\boldsymbol{x}-\boldsymbol{r}|)-H(|\boldsymbol{x}-\boldsymbol{r}|-|\varrho|)]^{\alpha}\boldsymbol{u}^{i}(t)+ \\ & +[H(|\boldsymbol{x}-\boldsymbol{r}|-|\varrho|)-H(|\boldsymbol{x}-\boldsymbol{r}|-|\varrho-\tfrac{1}{2}\delta|)]^{\alpha}\boldsymbol{u}^{s}(t) \end{aligned} \tag{4.116}$$

where it is understood that $(\boldsymbol{x}-\boldsymbol{r})$ is in the direction of $\boldsymbol{\delta}$ and in which the absolute value of the arguments contained in the Heaviside functions refer to quantities as indicated in Figs. 13 and 14 (a) for the deformed configuration of a single element or grain and its boundary zone. For the convenience of the reader the kinematics of this structural arrangement are shown again in Fig. 29 below. It is seen that ${}^{\alpha}\boldsymbol{u}^{i}(t)$ is the deformation of any arbitrary point within the microelement or grain and ${}^{\alpha}\boldsymbol{u}^{s}(t)$ the deformation arising from the surface interaction, whereby it is assumed that these deformation vectors are at least piecewise continuous functions of ${}^{\alpha}\boldsymbol{X}$ and t within a compact support, e.g. that the deformation fields are at least locally continuous. The quantity $H(|\cdot|)$ in relation (4.116) is the Heaviside function of the arguments in the brackets, which makes either one of the deformations zero valued depending upon whether one considers internal or surface deformations only.

From a systems theory point of view the function $H(|\cdot|)$ can be regarded as a "filtering operator". Hence one can also express the total deformation ${}^{\alpha}\boldsymbol{u}(t)$ as follows:

$$ {}^{\alpha}\boldsymbol{u}(t) = L_1{}^{\alpha}\boldsymbol{u}^i(t) + L_2{}^{\alpha}\boldsymbol{u}^s(t) \qquad (4.117) $$

in which the operators L_1 and L_2 are equivalent to those given in equations (2.32) and (4.116). For simplification of the present analysis each microelement is regarded as a continuum, where internal effects such as dislocations, for example, are neglected. Thus, on the basis of the internal microstress one can introduce as shown previously a surface stress such that:

$$ {}^{\alpha}\xi^s(t) = {}^{\alpha\beta}\boldsymbol{\tau}(t){}^{\alpha}\boldsymbol{n} \qquad (4.118) $$

in which ${}^{\alpha\beta}\boldsymbol{\tau}$ and ${}^{\alpha}\boldsymbol{n}$ have the meaning as given in Section 4.2. B. Furthermore, by following the representation given by Tonti[121] which has been used already in equation (4.45) one can define the internal microstrain of each element in a generalized form as follows:

$$ {}^{\alpha}\varepsilon^i_{ij}(t) = \tfrac{1}{2}(\delta_{jk}\nabla_i + \delta_{ik}\nabla_j){}^{\alpha}u^i_k(t) \qquad (4.119) $$

and correspondingly the microstrains valid on the surface of each microelement by:

$$ {}^{\alpha}\varepsilon^s_{ij}(t) = \tfrac{1}{2}(\delta_{jk}\nabla_i + \delta_{ik}\nabla_j){}^{\alpha}u^s_k(t) \qquad (4.120) $$

Again, as mentioned earlier, the gradient operation in (4.119) has to be performed with respect to the internal body coordinates, whilst that of equation (4.120) with respect to the surface coordinates. It is important to note that throughout the present formulation, the microstress ${}^{\alpha}\xi(t)$ is considered as a continuous quantity and the microstrain ${}^{\alpha}\boldsymbol{\varepsilon}(t)$ as a discontinuous one.

(B) *Operational formulation of relaxation phenomena*

It has been mentioned in the introduction that the present formulation considers the molecular relaxation process of crystalline solids to be ascribed to an internal relaxation of a single element and a surface relaxation to occur in the boundary zone between the elements. In attempting a unified formulation of the relaxation behaviour one can, to a first approximation, neglect the influence of the internal response to that of the grain boundary zone itself. In real crystalline solids, however, the presence of dislocations or other

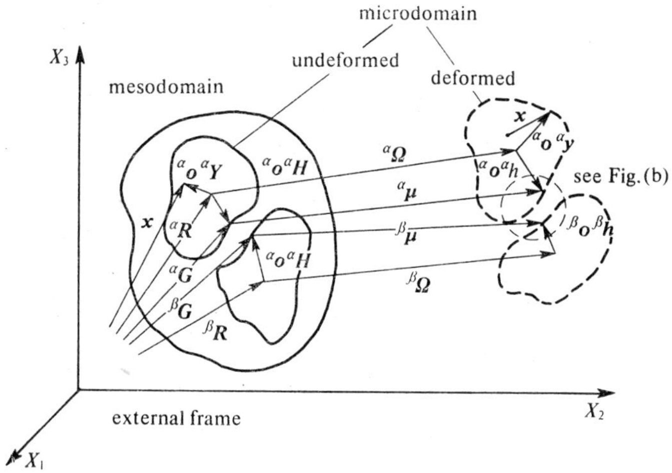

(a) microelement kinematics

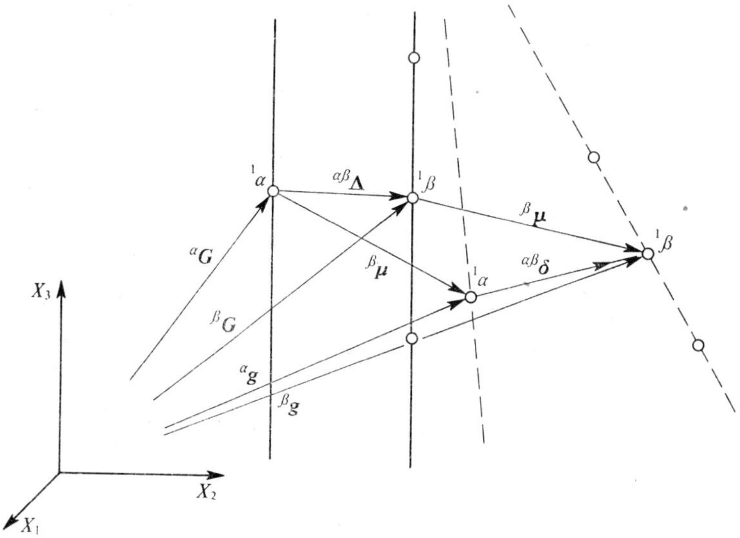

(b) grain boundary kinematics

Fig. 29.

disorder effects will influence the grain boundary behaviour and must therefore be taken into consideration.

In general, in order to express the stress-deformation or stress-strain relation for the internal relaxation of a microelement, a differential deformation law[104] can be used in the form of an mth order polynomial for the stress, and of an nth order polynomial for the strain, such that:

$$\begin{aligned} &{}^{\alpha}\boldsymbol{a}_0^i : \boldsymbol{\xi}^i(t) + {}^{\alpha}\boldsymbol{a}_1^i : \boldsymbol{\xi}^{i(1)}(t) + \ldots + {}^{\alpha}\boldsymbol{a}_m^i : \boldsymbol{\xi}^{i(m)}(t) \\ &= {}^{\alpha}\boldsymbol{b}_0^i : \boldsymbol{\varepsilon}^i(t) + {}^{\alpha}\boldsymbol{b}_1^i : \boldsymbol{\varepsilon}^{i(1)}(t) + \ldots + {}^{\alpha}\boldsymbol{b}_n^i : \boldsymbol{\varepsilon}^{i(n)}(t) \end{aligned} \tag{4.121}$$

in which the coefficients ${}^{\alpha}a_0^i$, ..., etc. are fourth order tensors. Analogously for the relaxation occurring in the grain boundary, i.e. within the zone in which an interaction potential between two adjacent grains is considered to exist, a similar differential law of deformation can be written as follows:

$$\begin{aligned} &{}^{\alpha}\boldsymbol{a}_0^s : \boldsymbol{\xi}^s(t) + {}^{\alpha}\boldsymbol{a}_1^s : \boldsymbol{\xi}^{s(1)}(t) + \ldots + {}^{\alpha}\boldsymbol{a}_1^s : \boldsymbol{\xi}^{s(m)}(t) \\ &= {}^{\alpha}\boldsymbol{b}_0^s : \boldsymbol{\varepsilon}^s(t) + {}^{\alpha}\boldsymbol{b}_1^s : \boldsymbol{\varepsilon}^{s(1)}(t) + \ldots + {}^{\alpha}\boldsymbol{b}_n^s : \boldsymbol{\varepsilon}^{s(n)}(t) \end{aligned} \tag{4.122}$$

The surface stresses within the boundary zone are considered to be continuous, whilst the microstrains are regarded as discontinuous and derivable from an interaction potential Ψ in a similar manner as shown for the elastic response of polycrystalline solids. This potential will be discussed in paragraph (iv) of this section. The conventional ways of obtaining solutions of equations (4.121) and (4.122) are the Laplace transform method or that of integral operators. However, in both these methods it is necessary to introduce "relaxation spectra" that correspond to the internal and surface relaxation, respectively.

(C) *Internal relaxation*

In order to consider the internal relaxation phenomenon of a single crystal α ($\alpha = 1, 2, \ldots, N$) it is necessary to assume that the coefficients ${}^{\alpha}\boldsymbol{a}$ and ${}^{\alpha}\boldsymbol{b}$ in relations (4.121) and (4.122) are independent of the microstress and microstrain. Following the operator formalism as shown in reference[104], it can be written for the present case that:

$$I^{*(r)}(x) = \underbrace{\int_0^t \ldots \int_0^t}_{r} x(\tau)\,\mathrm{d}\tau \tag{4.123}$$

where the superscript r designates the rth integral of the operator I^* and x a continuous function of time. Hence by multiplying this form of the operator with equation (4.121) one obtains:

$$\sum_{p=0}^{m} {}^{\alpha}\boldsymbol{a}^{i} I^{*(r-p)} : \xi^{i}(t) = \sum_{p=0}^{n} {}^{\alpha}\boldsymbol{b}^{i} I^{*(r-p)} : \varepsilon^{i}(t) \tag{4.124}$$

For simplification of the analysis, one can now write:

$$\begin{aligned} {}^{\alpha}\boldsymbol{A}^{i}(I^*) &= \sum_{p=0}^{m} {}^{\alpha}\boldsymbol{a}^{i} I^{*(r-p)} \\ {}^{\alpha}\boldsymbol{B}^{i}(I^*) &= \sum_{p=0}^{n} {}^{\alpha}\boldsymbol{b}^{i} I^{*(r-p)} \end{aligned} \tag{4.125}$$

where the operators ${}^{\alpha}\boldsymbol{A}^{i}$ and ${}^{\alpha}\boldsymbol{B}^{i}$ can be used to establish a constitutive relation for the mechanical relaxation of the individual microelement by substituting equation (4.125) into (4.124). This results in:

$$ {}^{\alpha}\xi^{i}(t) = [{}^{\alpha}\boldsymbol{A}^{i}(t)]^{-1} : ([{}^{\alpha}\boldsymbol{B}^{i}(t)] : {}^{\alpha}\varepsilon^{i}(t)) \tag{4.126}$$

Equivalently, as shown in reference[83], one can express the above relation as follows:

$$ {}^{\alpha}\xi^{i}(t) = \left[{}^{\alpha}A_{0}^{i} + \sum_{p=1}^{r} \frac{{}^{\alpha}A_{p}^{i}(t)}{1 + {}^{\alpha}\lambda_{p}^{i} I^*} \right] : {}^{\alpha}\varepsilon^{i}(t) \tag{4.127}$$

in which the quantity ${}^{\alpha}A_{0}^{i}$ corresponds to an instantaneous elastic tensor modulus of the element, ${}^{\alpha}A_{p}^{i}(t)$ a relaxation tensor modulus and the quantity $(1 + {}^{\alpha}\lambda_{p}^{i} I^*)$ is an integral operator that contains the internal relaxation time $1/{}^{\alpha}\lambda_{p}^{i}$ of the single grain. It is further convenient for the treatment of the internal relaxation to introduce the following operational quantities[104]:

$$\mathscr{E}^*(-{}^{\alpha}\lambda_{p}^{i})x(\tau) = \int_{0}^{t} \exp[-{}^{\alpha}\lambda_{p}^{i}(t-\tau)]x(\tau)\,\mathrm{d}\tau \tag{4.128}$$

It has been shown in reference[104] that $\mathscr{E}^*(-{}^{\alpha}\lambda_{p}^{i})$ is equivalent to the operational form:

$$\mathscr{E}^*(-{}^{\alpha}\lambda_{p}^{i}) = \frac{I^*}{1 + {}^{\alpha}\lambda_{p}^{i} I^*} \tag{4.129}$$

Using the above operational formalism permits to express the constitutive law of a microelement so far as the internal relaxation is concerned in the following manner:

$$
\begin{aligned}
{}^{\alpha}\xi^{i}(t) &= {}^{\alpha}A^{i}(t) : {}^{\alpha}\varepsilon^{i}(t) - \\
&- \sum_{p=1}^{r} \int_{0}^{t} {}^{\alpha}A_{p}^{i}\, {}^{\alpha}\lambda_{p}^{i} \exp[-{}^{\alpha}\lambda_{p}^{i}(t-\tau)] : {}^{\alpha}\varepsilon^{i}(\tau)\,d\tau
\end{aligned}
\tag{4.130}
$$

Noting that the operator ${}^{\alpha}A^{i}$ is given by:

$$
{}^{\alpha}A^{i} = \sum_{p=0}^{r} {}^{\alpha}A_{p}^{i} \tag{4.131}
$$

and further introducing a tensorial relaxation function of the form:

$$
{}^{\alpha}\Gamma^{i}(t-\tau) = \sum_{p=1}^{r} {}^{\alpha}A_{p}^{i}\, {}^{\alpha}\lambda_{p}^{i} \exp[-{}^{\alpha}\lambda_{p}^{i}(t-\tau)] \tag{4.132}
$$

leads finally to the stress relaxation function for a single microelement as follows:

$$
{}^{\alpha}\xi^{i}(t) = {}^{\alpha}A^{i} : {}^{\alpha}\varepsilon^{i}(t) - \int_{0}^{t} {}^{\alpha}\Gamma^{i}(t-\tau) : {}^{\alpha}\varepsilon^{i}(\tau)\,d\tau \tag{4.133}
$$

which corresponds to the conventional form of a stress-time function as obtained from a phenomenological approach. It is to be noted that the integral in (4.133) can also be expressed in terms of an operator ${}^{\alpha}\Gamma^{*i}(t)$ such that:

$$
{}^{\alpha}\Gamma^{*i}(t) : {}^{\alpha}\varepsilon^{i}(t) = \int_{0}^{t} {}^{\alpha}\Gamma^{i}(t-\tau) : {}^{\alpha}\varepsilon^{i}(\tau)\,d\tau \tag{4.134}
$$

In this form the internal relaxation of a grain is represented in terms of an experimentally accessible relaxation modulus and measurable relaxation times $1/{}^{\alpha}\lambda_{p}^{i}$ ($p = 1, \dots, r$).

(D) *Grain boundary relaxation*

The response characteristics with regard to the grain boundaries differ from those formulated for the internal relaxation. In general, the relaxation phenomenon due to the interaction effect between two microelements or grains will depend in accordance with the present theory on the "rate of change" of the significant relative displacement vector ${}^{\alpha\beta}\boldsymbol{d}$ introduced in the study of polycrystalline solids. This type of interaction has been ascribed to the existence of a potential function of the Morse type. However, without recourse to any specific potential form, a "generalized rate dependent potential" can be written as follows:

$$\Psi = \Psi(\boldsymbol{d}, \dot{\boldsymbol{d}}) \tag{4.135}$$

in which $\dot{\boldsymbol{d}}$ is the rate of change with time of the distance vector $\boldsymbol{d}$ employed in the stochastic deformation theory. Thus the generalized interaction force discussed earlier in this chapter as well as in Chapter II must be derived from this generalized potential by taking the total derivative of Ψ with respect to ${}^{\alpha\beta}\boldsymbol{d}$. In accordance with relation (4.118) the generalized interaction force becomes now:

$${}^{\alpha\beta}\boldsymbol{\tau}(t) = \frac{\partial\Psi}{\partial\boldsymbol{d}} + \frac{\partial\Psi}{\partial\dot{\boldsymbol{d}}}\,\mathrm{grad}\,\dot{\boldsymbol{d}} \tag{4.136}$$

so that the microstress within the grain boundary zone can be expressed by:

$${}^{\alpha}\xi^{s}(t) = \left(\frac{\partial\Psi}{\partial\boldsymbol{d}} + \frac{\partial\Psi}{\partial\dot{\boldsymbol{d}}}\,\mathrm{grad}\,\dot{\boldsymbol{d}}\right){}^{\alpha}\boldsymbol{n} \tag{4.137}$$

Similarly as before in order to deal with the grain boundary relaxation one can use a differential deformation law (4.121) and employ the integral operator formalism to obtain a relation between ${}^{\alpha}\xi^{s}(t)$ and ${}^{\alpha}\varepsilon^{s}(t)$ as follows:

$${}^{\alpha}\xi^{s}(t) = {}^{\alpha}\boldsymbol{A}^{s}:{}^{\alpha}\varepsilon^{s}(t) - \int_{0}^{t} {}^{\alpha}\boldsymbol{\Gamma}^{s}(t-\tau):{}^{\alpha}\varepsilon^{s}(\tau)\,\mathrm{d}\tau \tag{4.138}$$

in which the quantities ${}^{\alpha}\boldsymbol{A}^{s}$ and ${}^{\alpha}\boldsymbol{\Gamma}^{s}$ have the same meaning as before but are pertaining to the surface effect between two microelements. The right-hand side of equation (4.138) concerning the integral relating to the surface effect may also be written as follows:

$${}^{\alpha}\boldsymbol{\Gamma}^{*s}(t):{}^{\alpha}\varepsilon^{s}(t) = \int_{0}^{t} {}^{\alpha}\boldsymbol{\Gamma}^{s}(t-\tau):{}^{\alpha}\varepsilon^{s}(\tau)\,\mathrm{d}\tau \tag{4.139}$$

which is equivalent to the dissipative part of the generalized surface stress defined by relation (4.118).

In order to clarify these material characteristics, it is possible to formally introduce a "dissipative potential" in a generalized form by considering the potential as the inner product of the stress and strain or in terms of a quadratic form of the strain only (see also Moreau[67]). It can also be expressed as follows:

$$\Psi_D = \langle {}^{\alpha}\xi^{s}(t), {}^{\alpha}\varepsilon^{s}(t)\rangle \tag{4.140}$$

On basis of this definition of a dissipative potential, that reflects the rheological behaviour of the grain boundary, the former can also be expressed in a quadratic form by:

$$\Psi_D = \tfrac{1}{2}\{{}^{\alpha}A^{s} - {}^{\alpha}\Gamma^{*s}(t)\} :: {}^{\alpha}\varepsilon^{s}(t)\,{}^{\alpha}\varepsilon^{s}(t) \tag{4.141}$$

in which "::" means a fourth order contraction. It is seen that in contrast to the purely elastic case where the potential on basis of the coincidence cell theory depends on the relative displacement vector only, in the rheological formulation the potential becomes also dependent on the rate of change with time of this displacement vector. Using the previously defined generalized strains, the latter can also be written in terms of this vector as follows:

$${}^{\alpha}\varepsilon^{s}_{ij}(t) = \frac{1}{2}\left(\frac{\partial d_i}{\partial x_j} + \frac{\partial d_j}{\partial x_i}\right) \qquad (i, j = 1, 2, 3) \tag{4.142}$$

In order to obtain analytically the rheological characteristics of the grain boundary zone material, it becomes necessary to carry out an optimization procedure of the quantity $|\Psi_D - \Psi|$ with respect to some weighting function. This would then permit to establish a set of simultaneous linear differential equations from which the material parameters ${}^{\alpha}A^{s}$ and ${}^{\alpha}\Gamma^{s}$ could be evaluated. This type of optimization procedure has been indicated earlier in dealing with the response behaviour of fibrous systems. In certain cases it is possible to establish these parameters experimentally from spectroscopic studies.

(E) *Macroscopic mechanical relaxation*

(a) *Generalized field quantities*

In the formulation of the mechanical relaxation of crystalline solids as presented in the foregoing subsections, it has been pointed out that two distinct quantities for the associated field variables must be considered. Thus to bring about an overall relaxation relation, these quantities must be generalized in the sense of Gel'fand and Shilov[15] as shown on several occasions in this text. Hence, one can write a "generalized microstress" as follows:

$$
\begin{aligned}
{}^{\alpha}\boldsymbol{\sigma}(t) &= [H(|\boldsymbol{x}-\boldsymbol{r}|)-H(|\boldsymbol{x}-\boldsymbol{r}|-|\boldsymbol{\varrho}|)]^{\alpha}\boldsymbol{\xi}^{i}(t)+\\
&+[H(|\boldsymbol{x}-\boldsymbol{r}|-|\boldsymbol{\varrho}|)-H(|\boldsymbol{x}-\boldsymbol{r}|-|\boldsymbol{\varrho}-\tfrac{1}{2}\boldsymbol{\delta}|)]^{\alpha}\boldsymbol{\xi}^{s}(t)
\end{aligned}
\tag{4.143}
$$

and by using the previous "filtering operators L_1 and L_2":

$$
{}^{\alpha}\boldsymbol{\sigma}(t) = L_1{}^{\alpha}\boldsymbol{\xi}^{i}(t)+L_2{}^{\alpha}\boldsymbol{\xi}^{s}(t) \tag{4.144}
$$

By employing the earlier equations (4.134) and (4.140) that correspond to the internal and surface relaxation behaviour of a single microelement, the generalized microstress can also be written as:

$$
\begin{aligned}
{}^{\alpha}\boldsymbol{\sigma}(t) &= L_1\{{}^{\alpha}\boldsymbol{A}^{i}-{}^{\alpha}\boldsymbol{\Gamma}^{*i}(t)\}:{}^{\alpha}\boldsymbol{\varepsilon}^{i}(t)+\\
&+L_2\{{}^{\alpha}\boldsymbol{A}^{s}-{}^{\alpha}\boldsymbol{\Gamma}^{*s}(t)\}:{}^{\alpha}\boldsymbol{\varepsilon}^{s}(t)
\end{aligned}
\tag{4.145}
$$

in which the material parameters ${}^{\alpha}\boldsymbol{\Gamma}^{*i}(t)$ and ${}^{\alpha}\boldsymbol{\Gamma}^{*s}(t)$ have been defined previously. It is of interest to note that if creep properties of the crystalline solid are to be investigated, a "generalized microstrain" would have to be used. This can be done in a completely analogous manner by using the material characteristics known as creep compliance ${}^{\alpha}\boldsymbol{C}$ and the creep function ${}^{\alpha}\boldsymbol{F}$ for each microelement so that the internal and surface strains become:

$$
\begin{aligned}
{}^{\alpha}\boldsymbol{\varepsilon}^{i}(t) &= {}^{\alpha}\boldsymbol{C}^{i}:{}^{\alpha}\boldsymbol{\xi}^{i}(t)-\int_0^t {}^{\alpha}\boldsymbol{F}^{i}(t-\tau):{}^{\alpha}\boldsymbol{\xi}^{i}(\tau)\,\mathrm{d}\tau\\
{}^{\alpha}\boldsymbol{\varepsilon}^{s}(t) &= {}^{\alpha}\boldsymbol{C}^{s}:{}^{\alpha}\boldsymbol{\xi}^{s}(t)-\int_0^t {}^{\alpha}\boldsymbol{F}^{s}(t-\tau):{}^{\alpha}\boldsymbol{\xi}^{s}(\tau)\,\mathrm{d}\tau
\end{aligned}
\tag{4.146}
$$

from which the "generalized microstrain" can be formed as follows:

$$
{}^{\alpha}\boldsymbol{E}(t) = L_1{}^{\alpha}\boldsymbol{\varepsilon}^{i}(t)+L_2{}^{\alpha}\boldsymbol{\varepsilon}^{s}(t) \tag{4.147}
$$

For the present considerations, only the generalized stress ${}^{\alpha}\boldsymbol{\sigma}(t)$ and the operators ${}^{\alpha}\boldsymbol{A}$ and ${}^{\alpha}\boldsymbol{\Gamma}^{*}$ will be used. In order to achieve a formulation which links the individual microscopic response to the macroscopic one it is convenient to use single operators of the form:

$$
\begin{aligned}
{}^{\alpha}\boldsymbol{\Lambda}^{i}(t) &= {}^{\alpha}\boldsymbol{A}^{i}-{}^{\alpha}\boldsymbol{\Gamma}^{*i}(t)\\
{}^{\alpha}\boldsymbol{\Lambda}^{s}(t) &= {}^{\alpha}\boldsymbol{A}^{s}-{}^{\alpha}\boldsymbol{\Gamma}^{*s}(t)
\end{aligned}
\tag{4.148}
$$

by means of which the generalized stress can be expressed in a simpler form as follows:

$$^{\alpha}\sigma(t) = L_1{}^{\alpha}\Lambda^i(t):{}^{\alpha}\varepsilon^i(t)+L_2{}^{\alpha}\Lambda^s(t):{}^{\alpha}\varepsilon^s(t) \tag{4.149}$$

in which the quantities $^{\alpha}\Lambda^i$ and $^{\alpha}\Lambda^s$ are characteristic of the internal and surface relaxation of a single microelement or grain, respectively.

(b) *The material operator*

It has been shown in previous applications of probabilistic micromechanics, that the material functional or operator connects stress and deformations in the macroscopic sense from the knowledge of the microscopic relations. It can be expressed in general by:

$$\mathcal{M} = \mathcal{M}\{\mathcal{A}, \mathcal{B}, \mathcal{P}; t, \theta\} \tag{4.150}$$

in which $\mathcal{A}$ and $\mathcal{B}$ are stochastic integro-differential operators representative of an ensemble of microelements and valid over a specific mesodomain and for a specific crystalline solid, $\mathcal{P}$ is the distribution function of the significant field variables involved in the relaxation process, t the time and θ the temperature. It should be noted in what follows that the two stochastic integro-differential operators $\mathcal{A}$ and $\mathcal{B}$ are constructed from the operators $^{\alpha}\Lambda^i$ and $^{\alpha}\Lambda^s$ by means of the distribution function $\mathcal{P}$. For simplicity of the present analysis, isothermal conditions are implied, which imposes the restriction on the model for the mechanical relaxation of crystalline solid in that the material operator is regarded as independent of the temperature. This will be valid in general for materials stressed under moderate temperatures only.

Referring to the analysis of crystalline solids given in Section 4.2 of this chapter, and which was concerned with the elastic response of such materials, it has been shown that the most important distribution function in the formulation is that of the orientation $\boldsymbol{O}$ of the single crystal. It has also been shown that the misfit angles of individual grain boundaries induce an auxiliary variable to this orientation. In view of these remarks one can express the stress valid within a particular mesodomain of the material sample in the following form:

$$^{M}\sigma(t) = \langle\sigma(t)\rangle = \sum {}^{\alpha}\sigma(t)\Delta\mathcal{P}\{\boldsymbol{O}, \varphi; t\} \tag{4.151}$$

where $\mathcal{P}\{\boldsymbol{O}, \varphi; t\}$ designates the joint distribution function of the random orientation $\boldsymbol{O}$ and the misfit angle φ. The second moment of ${}^{\alpha}\boldsymbol{\sigma}(t)$ can therefore be written as:

$$V_{\sigma}(t) = \sum_{i} \{{}^{\alpha}\boldsymbol{\sigma}(t) - {}^{M}\boldsymbol{\sigma}(t)\}^{2} \Delta\mathscr{P}\{\boldsymbol{O}, \varphi; t\} \tag{4.152}$$

An analogous expression for the mesoscopic strain cannot be written immediately. Such an expression can only be arrived at by consideration of the inter-relation between the operators linking the microstrain to the microstress, e.g. the creep function and creep compliance as well as the previously introduced filtering operators. For the present purpose, however, denoting the mesoscopic strain by ${}^{M}\boldsymbol{e}(t)$ which corresponds to ${}^{M}\boldsymbol{\sigma}(t)$ and using the material operator for the mesodomain yields:

$$ {}^{M}\boldsymbol{\sigma}(t) = \mathscr{M}(t) : {}^{M}\boldsymbol{e}(t) \tag{4.153}$$

Considering the constitutive relation for a microelement or crystal in its generalized form, i.e. equation (4.149) and the definition of the operator given by (4.150), the two stochastic integro-differential operators, $\mathscr{A}$, $\mathscr{B}$ can be formally expressed by:

$$\begin{aligned} \mathscr{A} &= \mathscr{A}[{}^{\alpha}\boldsymbol{\Lambda}^{i}(t), \mathscr{P}\{\ldots\}] \\ \mathscr{B} &= \mathscr{B}[{}^{\alpha}\boldsymbol{\Lambda}^{s}(t), \mathscr{P}\{\ldots\}] \end{aligned} \tag{4.154}$$

from which the material operator in terms of the characteristic operators ${}^{\alpha}\boldsymbol{\Lambda}^{i}(t)$ and ${}^{\alpha}\boldsymbol{\Lambda}^{s}(t)$ as well as the distribution function $\mathscr{P}\{\cdot\}$ becomes:

$$\mathscr{M} = \mathscr{M}\{{}^{\alpha}\boldsymbol{\Lambda}^{i}(t), {}^{\alpha}\boldsymbol{\Lambda}^{s}(t); \mathscr{P}\{\cdot\}, t\} \tag{4.155}$$

(c) *Macroscopic response relations*

The macroscopic response relations can now be formulated by identifying the macroscopic stress acting on the material sample with the time-dependent mesoscopic stress in terms of the previously discussed operators. Thus

$$ {}^{M}\boldsymbol{\sigma}(t) = \langle\boldsymbol{\sigma}(t)\rangle = \langle L_1 {}^{\alpha}\boldsymbol{\Lambda}^{i}(t) : {}^{\alpha}\boldsymbol{\varepsilon}^{i}(t)\rangle + \langle L_2 {}^{\alpha}\boldsymbol{\Lambda}^{s}(t) : {}^{\alpha}\boldsymbol{\varepsilon}^{s}(t)\rangle \tag{4.156}$$

However, the above summation procedure involves considerations of the individual crystal radius ϱ as well as the grain boundary displacement $\frac{1}{2}\boldsymbol{\delta}$ discussed earlier. Furthermore, on the assumption that the material operators ${}^{\alpha}\boldsymbol{\Lambda}^{i}(t)$ and ${}^{\alpha}\boldsymbol{\Lambda}^{s}(t)$ are statistically independent of the microstrains ${}^{\alpha}\boldsymbol{\varepsilon}^{i}(t)$ and ${}^{\alpha}\boldsymbol{\varepsilon}^{s}(t)$, respectively, (4.156) can be expressed in a reduced form as follows:

$$ {}^{M}\boldsymbol{\sigma}(t) = \langle\boldsymbol{\Lambda}^{i}(t)\rangle : \langle L_1 {}^{\alpha}\boldsymbol{\varepsilon}^{i}(t)\rangle + \langle\boldsymbol{\Lambda}^{s}(t)\rangle : \langle L_2 {}^{\alpha}\boldsymbol{\varepsilon}^{s}(t)\rangle \tag{4.157}$$

and by using the distribution function $\mathscr{P}\{\boldsymbol{O}, \varphi; t\}$ it becomes finally:

$$
\begin{aligned}
{}^{M}\boldsymbol{\sigma}(t) &= \langle \boldsymbol{A}^{i}(t)\rangle : \sum L_1{}^{\alpha} \boldsymbol{\varepsilon}^{i}(t) \Delta \mathscr{P}\{\boldsymbol{O}, \varphi; t\} + \\
&+ \langle \boldsymbol{A}^{s}(t)\rangle : \sum L_2{}^{\alpha} \boldsymbol{\varepsilon}^{s}(t) \Delta \mathscr{P}\{\boldsymbol{O}, \varphi; t\} \qquad (4.158)
\end{aligned}
$$

In the above formulation the joint distribution function $\mathscr{P}\{\boldsymbol{O}, \varphi; t\}$ has been used. However, at the beginning of the present analysis it has been stated the relaxation model of a crystalline solid is visualized as one in which the relaxation interior to each microelement differs from that of the grain boundary zone. As a consequence, it may be assumed as in the case of the elastic response of a crystalline solid (Section 4.2), that in the case of relaxation an analogous characteristic function will exist such that the mesoscopic strain will be related to either the internal or the grain boundary strain in the following manner:

$$
\begin{aligned}
(1-K^{r})\,{}^{M}\boldsymbol{e}(t) &= \sum L_1{}^{\alpha} \boldsymbol{\varepsilon}^{i}(t) \Delta \mathscr{P}\{\boldsymbol{O}, \varphi; t\} \\
K^{r}\,{}^{M}\boldsymbol{e}(t) &= \sum L_2{}^{\alpha} \boldsymbol{\varepsilon}^{s}(t) \Delta \mathscr{P}\{\boldsymbol{O}, \varphi; t\} \qquad (4.159)
\end{aligned}
$$

Thus equation (4.158) in terms of the above characteristic function K^{r} can be rewritten to give the mesoscopic stress-strain relations as follows:

$$
{}^{M}\boldsymbol{\sigma}(t) = [\langle \boldsymbol{A}^{i}(t)\rangle (1-K^{r}) + \langle \boldsymbol{A}^{s}(t)\rangle K^{r}] : {}^{M}\boldsymbol{e}(t) \qquad (4.160)
$$

By comparing (4.160) with (4.153) it is seen that the overall material operator ${}^{M}\mathscr{M}(t)$ replacing the conventional constitutive relations contains the "microelement operators" as well as the rather complicated function K^{r} which is associated with the disorder effects in the grain boundary during the relaxation process. In the above analysis a purely formal representation of the mechanical relaxation of crystalline solids has been given. Comprehensive studies of a quantitative nature are still required in order to obtain more information on the grain boundary relaxation and the disorder effect briefly introduced in the above theory. Such studies will be indicated in the following chapter of this monograph.

V. Experimental Micromechanics

5.1 Introduction

The concepts and mathematical aspects of the probabilistic micromechanics of structured media have been considered in the preceding chapters of this monograph. On the basis of these concepts a general theory of stochastic deformation has been developed and the application of the theory within the scope of this text has been given for two groups of structured solids in the foregoing chapter. The experimental work that has been carried out as well as other newer techniques at present under investigation in order to determine the characteristic quantities or their corresponding distribution functions as required by the developed theory, are the subject matter of this chapter. Again as pointed out earlier, the experimental methods discussed here and the achieved results concern the two groups of structured solids, i.e. polycrystalline solids and fibrous structures.

In general, experimental techniques of structured materials include amongst others transmission, field-ion and scanning electron microscopy, X-ray diffraction, X-ray and optical microscopy, spectroscopic methods, etc. Some of these techniques have been used to establish local effects and changes within single microelements and on boundaries of such elements[122–125]. However, such techniques have been employed to investigate usually a unique phenomenon on a carefully prepared specimen representing a single element of the microstructure. Consequently these techniques are either not immediately applicable or altogether inadequate for studies of structured solids in the sense of the micromechanics theory. In accordance with the latter theory it becomes necessary in the study of deformations to employ new techniques that permit measurements of an ensemble of microelements from which the associated statistics can be evaluated. One such technique is known as "stress-holographic interferometry", which in the case of crystalline solids is combined with X-ray diffraction to obtain microdeformations and microrotations of single crystals embedded in an appropriate matrix as discussed later. The holographic interferometry technique has been extended more recently to the measurement of the creep behaviour of fibrous systems under specified environmental conditions. This latter method, due to the limited scope of the present text will not be treated here, but is available in another publication[126]. It should be mentioned, however, that a brief discussion on the measurement of the distribution of the relative displacement vector ${}^{\alpha\beta}\boldsymbol{d}$ for such materials by means of scanning electron microscopy using electron back-scattering and cathodoluminescence will be given at the end of this chapter.

5.2 Evaluation of basic field quantities

It has been mentioned throughout this text that certain field quantities or their distribution functions over a particular mesodomain are experimentally accessible. In the case of polycrystalline solids the main kinematic quantities are the microdeformations and microrotations, which can be determined from their corresponding distribution functions. For fibrous systems, analogously, microdeformations and the angular distribution of fibre orientations in a two-dimensional network can also be determined.

In general, the stochastic relations given in Chapter II (equations (2.14), (2.15), (2.17) and (2.18)) can be considered so that at all times during the deformation process the kinematic relations will be as follows:

$$ {}^{\alpha}\boldsymbol{x}({}^{\alpha}\boldsymbol{X}; t) = {}^{\alpha}\boldsymbol{o}({}^{\alpha}\boldsymbol{O}, {}^{\alpha}\boldsymbol{Y}; t) \cdot {}^{\alpha}\boldsymbol{y}({}^{\alpha}\boldsymbol{Y}; t) + {}^{\alpha}\boldsymbol{r}({}^{\alpha}\boldsymbol{R}; t) \tag{5.1} $$

where the symbols have the meaning given in Section 2.4 (Chapter II). Thus, considering the expressions for the stochastic deformations of crystalline solids (equation (2.17)) and that of fibrous systems (equations (2.27) and (2.28)), the deformation of a microelement within the structure can be written as:

$$ {}^{\alpha}\boldsymbol{u}({}^{\alpha}\boldsymbol{R}; t) = {}^{\alpha}\boldsymbol{r}({}^{\alpha}\boldsymbol{R}; t) - {}^{\alpha}\boldsymbol{R} \tag{5.2} $$

So far, as the experimental determination of these basic quantities is concerned, the following statements can be made:

(i) All initial values, i.e. ${}^{\alpha}\boldsymbol{X}, {}^{\alpha}\boldsymbol{O}, {}^{\alpha}\boldsymbol{Y}, {}^{\alpha}\boldsymbol{R}$ are experimentally accessible and can be obtained from measurements of their corresponding distribution functions from various experimental techniques.

(ii) According to the developed theory ${}^{\alpha}\boldsymbol{o}({}^{\alpha}\boldsymbol{O}, {}^{\alpha}\boldsymbol{Y}; t)$ and ${}^{\alpha}\boldsymbol{y}({}^{\alpha}\boldsymbol{Y}; t)$ are stochastic processes. This is also the case for the function representing the deformed position vector ${}^{\alpha}\boldsymbol{r}({}^{\alpha}\boldsymbol{R}; t)$. Thus, for the experimental observation of these quantities, the evolution of the related distributions with time as well as second moments must be used.

It should be pointed out that the experimental techniques discussed subsequently refer only to two specific configurations of the structured materials, i.e. the undeformed and deformed state under the condition of a constant load and temperature and otherwise fixed specified environmental conditions. The study of the evolution of the stochastic deformation process itself would involve experimental measurements or estimates of the transition probabilities involved in such a process in accordance with the theoretical considerations given in Chapter III. So far, the required techniques to perform such measurements are at present being investigated, but are as yet not available.

It is convenient for the experimental observation of microdeformations of a polycrystalline solid, for example, to decompose the trajectory of ${}^{\alpha}\boldsymbol{r}$ (equation (5.2)) into its expected value $\langle {}^{\alpha}\boldsymbol{r} \rangle$ and the fluctuating part ${}^{\alpha}\boldsymbol{r}^{*}$ so that:

$$ {}^{\alpha}\boldsymbol{r} = \langle {}^{\alpha}\boldsymbol{r} \rangle + {}^{\alpha}\boldsymbol{r}^{*} \tag{5.3} $$

from which it follows by substituting into (5.2) that:

$$ {}^{\alpha}\boldsymbol{u} = \langle {}^{\alpha}\boldsymbol{u} \rangle + {}^{\alpha}\boldsymbol{u}^{*} = \langle \boldsymbol{r} \rangle - {}^{\alpha}\boldsymbol{R} + {}^{\alpha}\boldsymbol{r}^{*} \tag{5.4} $$

$$ \langle \boldsymbol{u} \rangle = \langle \boldsymbol{r} \rangle - {}^{\alpha}\boldsymbol{R}, \qquad {}^{\alpha}\boldsymbol{u}^{*} = {}^{\alpha}\boldsymbol{r}^{*} \tag{5.5} $$

Hence, the stochastic deformation of an arbitrary point of the αth microelement or grain will become:

$$ {}^{\alpha}\boldsymbol{w}({}^{\alpha}\boldsymbol{X}; t) = {}^{\alpha}\boldsymbol{x}({}^{\alpha}\boldsymbol{X}; t) - {}^{\alpha}\boldsymbol{X} = {}^{\alpha}\boldsymbol{o}\,{}^{\alpha}\boldsymbol{y} - {}^{\alpha}\boldsymbol{O}\,{}^{\alpha}\boldsymbol{Y} + {}^{\alpha}\boldsymbol{u} \tag{5.6} $$

Similarly one can decompose the rotation tensor ${}^{\alpha}\boldsymbol{O}$ and the position vector ${}^{\alpha}\boldsymbol{Y}$ so that the total microdeformation ${}^{\alpha}\boldsymbol{w}$ is given by:

$$\begin{aligned} &{}^{\alpha}w({}^{\alpha}X;\,t) = \langle w \rangle + {}^{\alpha}w^* && \text{(a)} \\ &\langle w \rangle = \langle o \cdot y \rangle - {}^{\alpha}O^{\alpha}Y + \langle u \rangle && \text{(b)} \\ &{}^{\alpha}w^* = {}^{\alpha}o \cdot {}^{\alpha}y^* + {}^{\alpha}o^* \cdot {}^{\alpha}y + {}^{\alpha}o^* \cdot {}^{\alpha}y^* + {}^{\alpha}u^* && \text{(c)} \end{aligned} \qquad (5.7)$$

Thus the distribution function of the total deformation ${}^{\alpha}\boldsymbol{w}$ can be determined uniquely from the knowledge of the distributions of ${}^{\alpha}\boldsymbol{u}, {}^{\alpha}\boldsymbol{o}$ and ${}^{\alpha}\boldsymbol{y}$ at a specific time. Hence, for a model representing a polycrystalline solid the values of the quantities given by relation (5.4) can be determined by using "stress-holographic interferometry". Similarly by means of X-ray diffraction studies the change of the crystallographic orientation, i.e.:

$$({}^{\alpha}o - {}^{\alpha}O) = \langle o - O \rangle + ({}^{\alpha}o - {}^{\alpha}O)^* \qquad (5.8)$$

can also be determined.

It should be noted that the quantities ${}^{\alpha}\boldsymbol{u}$ and ${}^{\alpha}\boldsymbol{o}$ or their fluctuations from the mean values, i.e. ${}^{\alpha}\boldsymbol{u}^*$ and ${}^{\alpha}\boldsymbol{o}^*$ may assume one of several values during the deformation of the material. Thus, $\boldsymbol{u}^*$, for example, can be represented by its three components u_1^*, u_2^* and u_3^* in a Euclidean space over the range in which their respective probability functions can be defined. Similar considerations hold for the rotations, which could be represented in terms of polar coordinates ν and $\varkappa$ of the axis of rotation together with the angle of rotation η. Alternatively, if the rotation is representable by an orthogonal matrix such as ${}^{\alpha}\boldsymbol{o}^*$ any three elements of this matrix, which are independent, can be taken as the components of rotation. Whilst the microdeformations and microrotations are fully determined by their components in a Euclidean space, they are equally representable in the probability space X or from an experimental point of view, in the deformation space U by a set of discrete observations. Thus the number of discrete observations N in general can be regarded as to coincide with the number of microelements ($\alpha = 1, 2, \ldots, N$) which are contained within a specified mesodomain of the material sample. The class of subsets belonging to the probability space X as discussed in Chapter III, contains the following sets, written in terms of the measurable microdeformations:

$$\begin{aligned} &U = \{{}^{k}\eta_i \leqslant {}^{\alpha}u_i \leqslant {}^{k}\eta_i + \Delta^{k}\eta_i\} \\ &i = 1, 2, 3, \quad \alpha = 1, 2, \ldots, N, \quad k = 1, 2, \ldots, M \end{aligned} \qquad (5.9)$$

where the superscript k ($k = 1, \ldots, M$) designates specific values of the measurable microdeformations. According to the experimental procedure discussed subsequently involving the scanning of holograms, two observations along individual scanning lines are assessed with an accuracy of $\Delta^k\eta_i$. Hence the new set of observations can be expressed by:

$$^k\eta_i + \Delta^k\eta_i = {}^{k+1}\eta_i \tag{5.10}$$

where $^k\eta_i \in Z^+$. With reference to the previously used probability measures (Chapters II and III) it is evident that:

$$\mathscr{P}\{^k\eta_i \leqslant {}^\alpha u_i \leqslant {}^k\eta_i + \Delta^k\eta_i\} = \mathscr{P}\{^\alpha u_i \in U\} \tag{5.11}$$

Considering now a subclass U' of U then the "joint probability distribution function" will be given by:

$$\mathscr{P}\{U'\} = \sum_{^\alpha u_i \in U'} p(u_i), \quad \text{where } U' = \{^\alpha u_i < {}^k\eta_i\} \tag{5.12}$$

and in which $p(^\alpha u_i)$ is the density of the probability distribution. In an alternative form

$$\mathscr{P}\{^\alpha u_1 < {}^r\eta_1, {}^\alpha u_2 < {}^s\eta_2, {}^\alpha u_3 < {}^t\eta_3\}$$
$$= \sum_{^\alpha u_1 < {}^r\eta_1} \sum_{^\alpha u_2 < {}^s\eta_2} \sum_{^\alpha u_3 < {}^t\eta_3} p(^\alpha u_1, {}^\alpha u_2, {}^\alpha u_3), \quad r, s, t < M \tag{5.13}$$

The above joint probability distribution function describes, in general, the distribution of microdeformations or rotations. However, for a more precise characterization of these random quantities, as obtained from experimental observations, one can also employ "marginal distributions". As shown earlier in Chapter I, the marginal distribution for the one-dimensional vector component "u_1" of the deformation, can be written for instance as follows:

$$\mathscr{P}\{u_1 < {}^r\eta_1\} = \sum_{u_1 < {}^r\eta_1} \sum_{u_2 < \eta_2} \sum_{u_3 < \eta_3} p(u_1, u_2, u_3) \tag{5.14}$$

or

$$\mathscr{P}\{u_1 < {}^r\eta_1\} = \sum_{u_1 \in U'} \sum_{u_2, u_3 \in U} p(u_1, u_2, u_3) \tag{5.15}$$

One can similarly define the distributions for the other two vector components of the microdeformation. In particular, if the random variables are assumed to be statistically independent, the above form reduces to:

$$\mathscr{P}\{u_i < \eta_i;\ i = 1, 2, 3\}$$
$$= \mathscr{P}\{u_1 < \eta_1\}\mathscr{P}\{u_2 < \eta_2\}\mathscr{P}\{u_3 < \eta_3\} \tag{5.16}$$

In an analogous manner, expressions can be written for the expected values of the three vector components as well as their mixed moments about the origin, as discussed in Chapter I, and thus a full description of these quantities can be obtained from experimental observations. It should be noted, however, that the distributions are usually obtained from a finite number of measurements and must be fitted for all practical purposes with a distribution law that is valid for a large number of measurements. As a consequence the nearly Gaussian character of the distribution functions for microdeformations and microrotations for a crystalline solid as presented subsequently, were obtained from the experimental data by a "goodness of fit" method, that corresponds to the various tests performed on the material sample. The above remarks were concerned with the general determination of the basic field quantities, whilst the actual test procedures to obtain the required information are described in the following sections.

5.3 Experimental investigations of crystalline solids

In order to obtain the distribution functions pertaining to microdeformations or microrotations in a crystalline solid a combined technique has been used, which consisted of the application of holographic interferometry and X-ray diffraction. It is the purpose of this section to briefly discuss both techniques and to present at the end some of the results which have been obtained by using a material model of a so-called "two-phase" structure consisting of aluminium monocrystals embedded in an epoxy-resin matrix.

(A) *Holographic interferometry*

Holographic interferometry became more recently one of the most important applications of the experimental technique known as "holography". This type of interferometry is basically concerned with the formation and interpretation of a "fringe pattern" that evolves, when adjacent light wave fronts interfere with each other during the so-called "reconstruction period" of a hologram. A detailed discussion on the theoretical background of holography and its application to interferometry has been given amongst others in references[127–129]. It has been discussed in these references how holograms produce an exact duplication of the light wave front which is reflected from an arbitrary object. In the application of this concept to interferometry so that measurements of deformations of structured solids can be achieved, the hologram is exposed twice. This permits during the reconstruction period in the test to observe "two objects" simultaneously. To clarify this further, if one "object" is considered to be the "undeformed state" of the material sample and the other object, corresponding to the "second exposure" of the holographic film, is associated with the "deformed state" of the sample, a simultaneous observation of the two "objects" makes small deformations observable due to the interference of the light wave front reflected from the two different configurations of the deformable solid. Hence this technique has been named "stress-holographic interferometry" or briefly the SHI-method.

If for the above observation of deformations one hologram only is constructed, the method is referred to as "single hologram technique". This method and the double exposure technique provides a fringe pattern which is fixed in time and the resulting fringes are also called "fixed fringes". Hence by employing this technique it becomes necessary to "move the observation point" in order to obtain a "fringe reading", which in terms of the number of fringes observed, is directly proportional to the occurring deformation. For a more detailed study of possible methods of analysis of interference fringes the reader is referred to references[130–133]. Without going into detail of the employed holographic interferometry technique a brief discussion on the single hologram-double exposure technique concerning the measurement of microdeformations of crystalline solids is given here. Thus the holographic arrangement for this technique used on models of crystalline solids, as mentioned earlier, is indicated in Fig. 30, and an overall view of the arrangement together with that for X-ray diffraction measurements is given in Fig. 31.

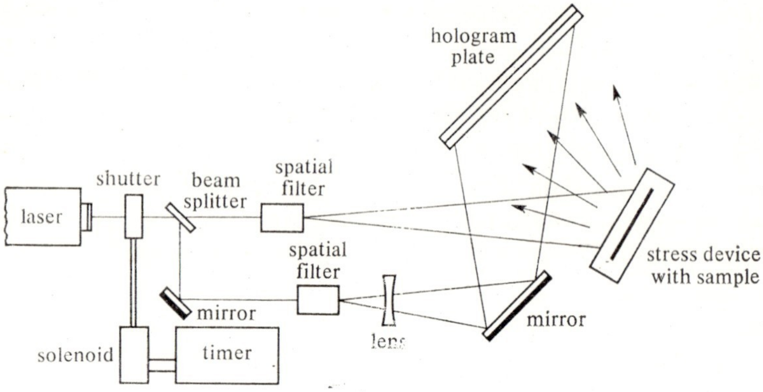

Fig. 30. Schematic lay-out of holographic interferometry technique.

Fig. 31. Holographic interferometry and X-ray diffraction assembly.

Fig. 30 further shows schematically the stress device in which the material sample is carefully mounted and subjected to the required loading. The position is also that of the virtual image during the second step in the test procedure, i.e. the reconstruction stage of the holographic interferometry. It is seen that the illuminating beam using a laser source emanates from a spatial filter (source *S*) for the experiments on models of crystalline solids. In conjunction with the continuous laser source (He–Ne gas laser) a shutter with a solenoid and timing device has been used to permit the exact timing of the double exposure required in the experiment. The optical arrangement contains further a beam splitter, mirrors and spatial filters, a supporting frame for the hologram plate as well as supports for other optical components used in the test. All components are mounted on magnetic bases which supply a firm contact with the massive holographic supporting table. The latter consists of a heavy concrete slab mounted on specially designed air springs so as to obtain a vibration-free setting required for the performance of high accuracy interferometry tests. For the testing of crystalline solids an automatically driven telescope read-out instrument was used, which was accurately focussed on particular points of the holograms and permitted the observation of a great number of fringe readings to an accuracy of $\pm 1/4$ of a fringe. A slightly more advanced method by means of which an "automatic scanning" of the hologram can be carried out to an accuracy of $\pm 1/8$ of a fringe during the observations will be described later in discussing the testing of fibrous structures.

In order to explain briefly the fundamental geometry underlying the holographic interferometry method, use will be made of Fig. 32 and Fig. 33. A more detailed study is given in reference[134]. Thus Fig. 32 indicates how the illuminating beam emanating from the spatial filter (1) meets the material sample surface. It is seen that the light beam is reflected from a surface point C^u in the undeformed state that corresponds to the point C^d in the deformed state of the material sample by the "object beam" onto the hologram plate. The positions of an "observer" during the reconstruction stage in the holography are indicated by the points O_i, O_j $(i, j = 1, 2, 3, 4)$. It is apparent from this geometry that the occurring deformation vector ${}^{\alpha}\boldsymbol{u}$ induces

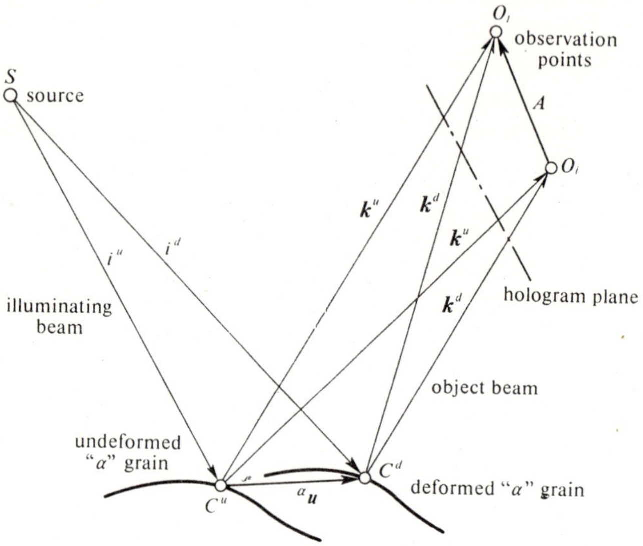

Fig. 32. Geometry of the general case of stress-holographic interferometry (SHI-method).

a phase-difference of the light beams. This phase difference is valid across the surface of the sample and can be expressed, in general[134], as follows:

$$\Delta_{ij} = [\boldsymbol{k}_i - \boldsymbol{k}_j]^{\alpha}\boldsymbol{u} = A^{\alpha}\boldsymbol{u} = 2\pi N_{ij} \tag{5.17}$$

in which Δ_{ij} denotes the phase difference of the light vectors $\boldsymbol{k}_i$, $\boldsymbol{k}_j$ in radians and N_{ij} the number of fringes that can be observed from the hologram when the observer moves from the position i to the position j. In general, for the case of four observation points $O_1 \to O_4$ there will be six different vectors, three of which may be chosen arbitrarily so as to form a non-orthonormal reference frame. A particular choice for such a reference frame is indicated in Fig. 33 below, which also shows the base vectors corresponding to the phase difference vectors and where for simplicity of the argument the light vectors for the undeformed and deformed states have been taken to be approximately equal, i.e. $\boldsymbol{k}_i^u = \boldsymbol{k}_i^d = \boldsymbol{k}_i$, etc.

In this context it should be mentioned that, for the performance of interferometry tests on fibrous networks with an equal accuracy

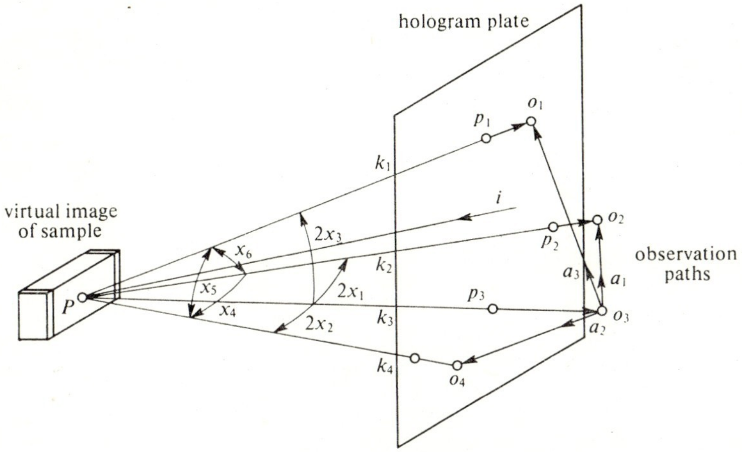

Fig. 33. Reconstruction geometry for a holographic interferogram.

of fringe readings in three directions, a different reconstruction geometry has to be adopted as discussed later. The microdeformations of individual crystals $^{\alpha}\boldsymbol{u}$ can be found by using the above reconstruction geometry, where in general:

$$^{\alpha}\boldsymbol{u} = u^j a_j \tag{5.18}$$

in which the a_j ($j = 1, 2, 3$) are the covariant base vectors and u^j the contravariant components of the microdeformation vector $^{\alpha}\boldsymbol{u}$. A detailed evaluation of the three-dimensional deformation for the discussed model is given in reference[134]. The magnitude $|^{\alpha}\boldsymbol{u}|$ can be obtained from the knowledge of the geometry, the wavelength of the light source, and the obtained fringe readings. It is important to note, however, that in performing the fringe readings by means of the telescopic read-out instrument a certain sign convention must be adopted. Thus, for example, if a "dark fringe" traverses the surface point under observation from point O_4, the fringe pattern will be divided into two sets. One can arbitrarily assign a positive or negative sign to either of these sets but the convention must be maintained throughout the read-out procedure of each test. For the purpose of illustrating the above described holographic interferometry technique to structured solids, a simple model was used in which the material sample consisted of aluminium monocrystals embedded in an epoxy-resin matrix. Typical stress holographic interferograms of this model indicating the fairly closed packed arrangement of the individual crystals and which was tested in tension with a load of 2.5 gm/mm^2 and 5 gm/mm^2, respectively are given in Figs. 34 and 35. It is clearly visible from these interferograms that the density of interference fringes within individual crystals changes with the applied stress level.

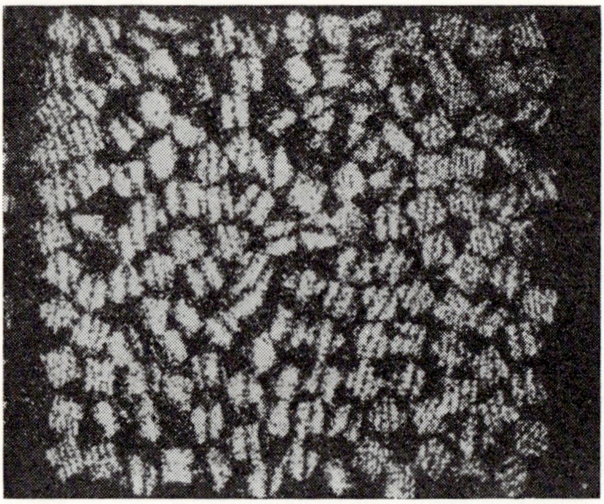

Fig. 34. Stress holographic interferogram of the model of a crystalline solid. Average stress field—2.5 gm/mm^2 (tension) (scale factor 3.5:1).

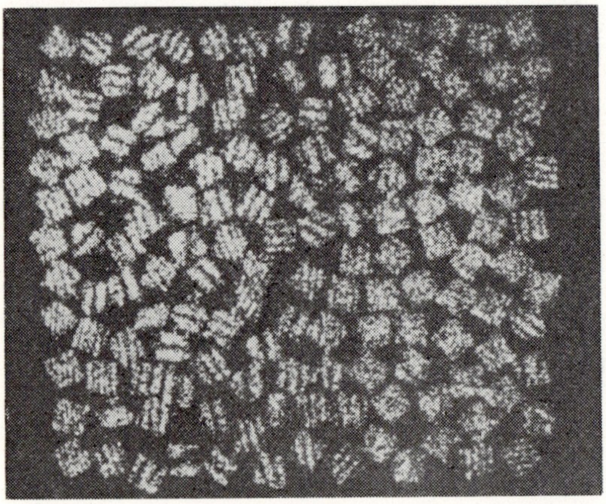

Fig. 35. Stress holographic interferogram of the model of a crystalline solid. Average stress field—5 gm/mm^2 (tension) (scale factor 3.5:1).

The aluminium crystals employed in the model were $2 \times 2 \times 2$ mm monocrystals of 99.995% purity and an orientation [0, 0, 1]. The precision of the latter was 10′. The initial dislocation density of the crystals was estimated to be less than 2.5×10^6 cm/cm^3. It should be mentioned that since the aluminium crystal structure is FCC with a lattice parameter $\boldsymbol{a} = 4{,}049$ Å at 25°C (see also reference[135]), it behaves in general anisotropically. However, the elastic modulus matrix can be considered to contain only three independent elastic coefficients. Hence from an experimental point of view, it becomes important to consider the effect of the so-called modulus ratio of these three constants in relation to the modulus of the epoxy-resin matrix. It was further evident that in order to avoid any plastic deformations of the crystals the stressing had to remain below the "critical resolved shear stress", e.g. the lowest shear stress at which the monocrystals begin to exhibit plastic deformations. In the employed model this critical shear stress was estimated to be of the order of 60 gm/mm^2 at room temperature as reported in reference[136]. Hence the magnitude of the applied stress levels in the above model was kept well below this value as well as considerations were given to the modulus ratio as mentioned above. In order to carry out the correct holographic interferometry on this type of the material model, the viscoelastic behaviour of the epoxy-resin matrix had also to be taken into account. The actual distributions of micro-deformations obtained by the application of the SHI-method will be shown subsequently.

(B) *SHI-method and X-ray diffraction*

As stated previously, in order to assess the microrotations of the elements of the structure, it is necessary, in addition to the SHI-method, to use an X-ray diffraction technique. For this purpose several X-ray diffraction methods have been studied and the back-reflection Laue method has been selected for the testing of models of crystalline solids. Before discussing this method in combination with the holographic interferometry, it may be instructive to mention briefly some general aspects of X-ray diffraction. A more detailed account of this technique is given in references[122,137].

The theory of X-ray diffraction is the same as that for the light diffraction already mentioned in the interferometry technique, except for the actual employed wavelength of the source. Thus, if a monochromatic parallel beam of light falls onto a "plane grating", it is reflected in directions for which the phase difference of scattered waves is equal to "$n\lambda$". In these directions the waves are in phase and reinforce each other. The integer "n" is referred to as the order of the spectrum. So far as the present model is concerned in which monocrystals are used, a simple representation of the "diffraction condition" is obtainable from Bragg's law[138], which is based on the notion of the reflection of X-rays from a "lattice plane" rather than on the diffraction of lattice rows. Similar conditions prevail in a three-dimensional array of atoms that forms the monocrystal. Thus considering a two-dimensional condition as is the case for the model (see test sample Fig. 40) under investigation, then with reference to Fig. 36 (a) for instance, one can write three conditions for the "constructive interference" such that:

$$\boldsymbol{r} \cdot \boldsymbol{G}_i - \boldsymbol{r}_0 \cdot \boldsymbol{G}_i = (\boldsymbol{r} - \boldsymbol{r}_0)\boldsymbol{G}_i = n_i \lambda \quad (i = 1, 2, 3) \tag{5.19}$$

These relations are referred to as the components of the Laue equation, in which each of them ($i = 1$, $i = 2$, $i = 3$) represents a condition for diffraction by a single row of lattice points. The set of integers n_1, n_2, n_3 is then the order of the spectrum and can be related to the well-known "Miller indices" employed in crystallography[50]. The simpler condition of diffraction known as Bragg's law employs the notion of the diffraction of X-rays from a lattice plane as a whole. This is indicated schematically in Fig. 36 (b). It is seen that the primary X-ray beam forms an angle θ with the crystal plane and the reflected beam can be defined as one which is composed of a large number of scattered rays mutually reinforcing one another. Such a reinforcement will occur only when the "path difference" of $DE + EF$ Fig. 36 (b) becomes equal to a whole number of wavelengths. This condition is expressed by Bragg's law as follows:

$$n\lambda = 2d\sin\theta \tag{5.20}$$

When this law is satisfied the diffracted waves from all lattice points of the crystal are in phase. It may be concluded therefore, that the diffraction of a crystal differs from that of a plane grating in that

as indicated by the "slit collimator" in the arrangement shown in Fig. 37 (b). By using this type of back-reflection technique 10–15 flag diffraction Laue pattern can be recorded simultaneously at a time. If the beam diameter and the grain size of the material sample are matched, one can expose along a certain scanning line of the sample (see also Fig. 41) a sufficient number of crystals to X-ray radiation and diffraction.

The actual geometry of the back-reflection Laue spot technique is illustrated by the following Fig. 38. The individual aluminium crystal is considered to be illuminated by an X-ray beam in which the beam

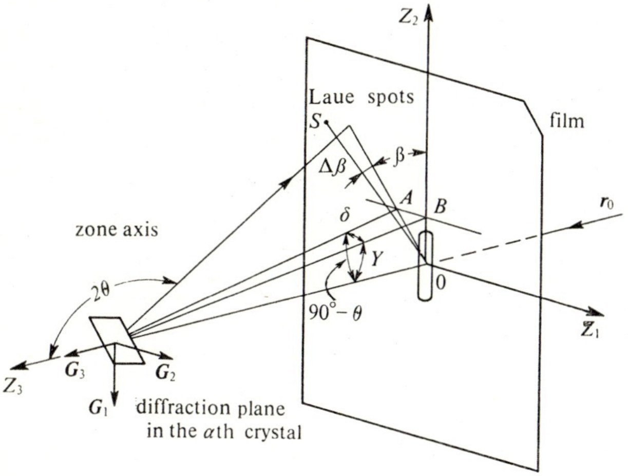

Fig. 38. Geometry of the back-reflection "Laue spot".

direction is $\boldsymbol{r}_0$ and which is identical to that of $\boldsymbol{G}_3$. The beam diffracted by the plane shown strikes the film at point "S", whilst the normal to the plane intersects the film at point "A". If the axes Z_1, Z_2, Z_3 are translated into point O, the definition of the position of the normal to the plane can be given in terms of the angles δ and γ. When Z_3 coincides with $\boldsymbol{G}_3$, the angles δ and γ can be determined from a "unit stereographic triangle" (see references[50,139]). The angle β and $180°-2\theta$ (Fig. 38) specify the position of the Laue spot in the plane of the film. It is to be noted that the angle β can be employed to determine the orientation of the αth crystal or grain with respect to the Z_2-axis, if the material model is considered as a two-dimensional one, as in the present investigation. The angle β and $180°-2\theta$ for a given diffraction spot are obtained in terms of the angles δ and γ as follows:

$$\tan\beta = \frac{\tan\delta}{\tan\gamma} \quad \text{(a)}$$
$$\tan(90°-\theta) = \frac{\tan\delta}{\sin\beta\cos\gamma} \quad \text{(b)} \qquad (5.23)$$

This relation holds for any arbitrary orientation of the crystal and hence can be used to determine the orientation of individual crystals. In the experimental study discussed here, the crystals were however "pre-oriented" in the sense that the stereographic projections could be directly related to the pattern of Laue spots observed on the film without recourse to the angles δ and γ (see reference[140]). Such a situation is indicated in Fig. 39 which shows the relation between the diffraction spot "S" and the stereographic projection "H" of the plane causing the spot (back-reflection). This type of back-reflection

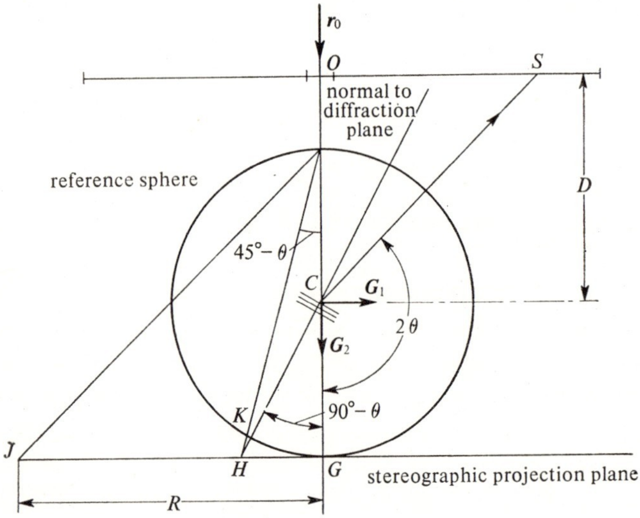

Fig. 39. Relation between the diffraction spot S and the stereographic projection H of the plane causing the spot (back-reflection).

relation has been employed in the combined SHI-method and X-ray diffraction experiment to identify the back-reflection Laue spots induced by the aluminium monocrystals of the material model. In this context some remarks to the latter should be made. Thus the model used was essentially a two-dimensional one in which the monocrystals were randomly distributed in the Z_1-Z_2 plane. A typical material sample is shown in Fig. 40, and an enlarged photograph of the test zone of such a specimen in Fig. 41.

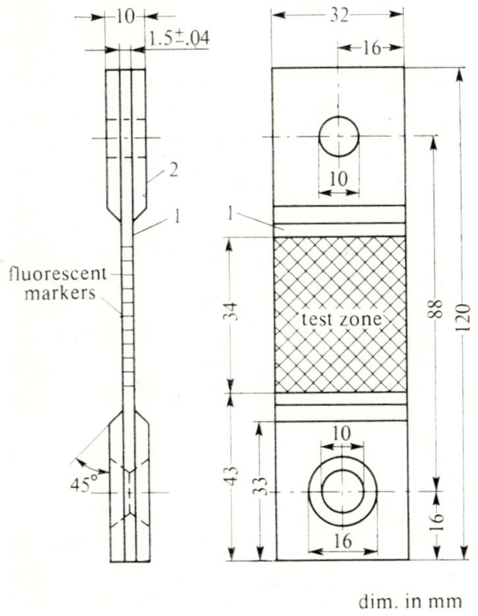

Fig. 40. Typical material sample for testing by the SHI- and X-ray diffraction method.

The application of the stress to this model occurred in the Z_1-direction and the microdeformations could be observed from one side of the sample in the manner described on pp. 189–190 of this section. The microrotations of the crystals could be observed from the other side of the material specimen. The testing arrangement for the combined methods of X-ray diffraction and holographic interferometry is shown schematically in Fig. 42. It is seen from this figure that on the left side of the stress application device containing the sample (Fig. 40) an X-ray diffraction Laue camera employing a slit collimator has been used. On the right side of the sketch, the holographic interferometry arrangement including the electro-optical fringe reading instrument is indicated. Since it has been mentioned earlier that the positions of the Laue spots on the X-ray film are governed by Bragg's law, it was desirable for the experimental investigation to vary the angle θ by varying the wavelength λ of the

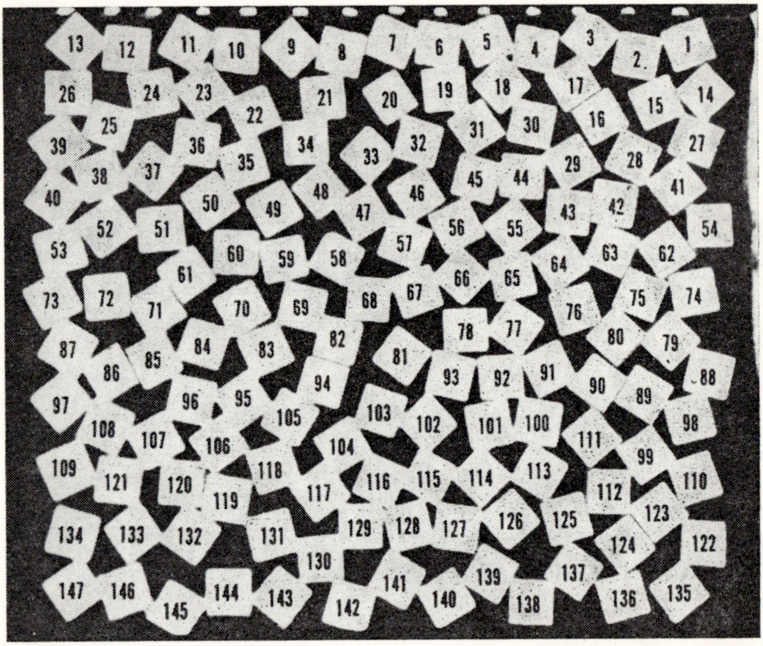

Fig. 41. Photograph of the "holographic side" of the model of the crystalline solid with the monocrystals and scanning lines (fluorescence markers).

source. Furthermore, by varying the focal spots of the X-ray source "scanning tests" on the specimen, as mentioned previously, could be performed. Hence, by using a "multifocus X-ray tube" with interchangeable target and cathode guns, the scanning of the specimens according to the scanning lines indicated in Fig. 41, which were clearly marked by fluorescence markers, could be carried out. In this context, it is of interest to note that the actual scanning has been carried out by moving the material sample parallel to the plane of

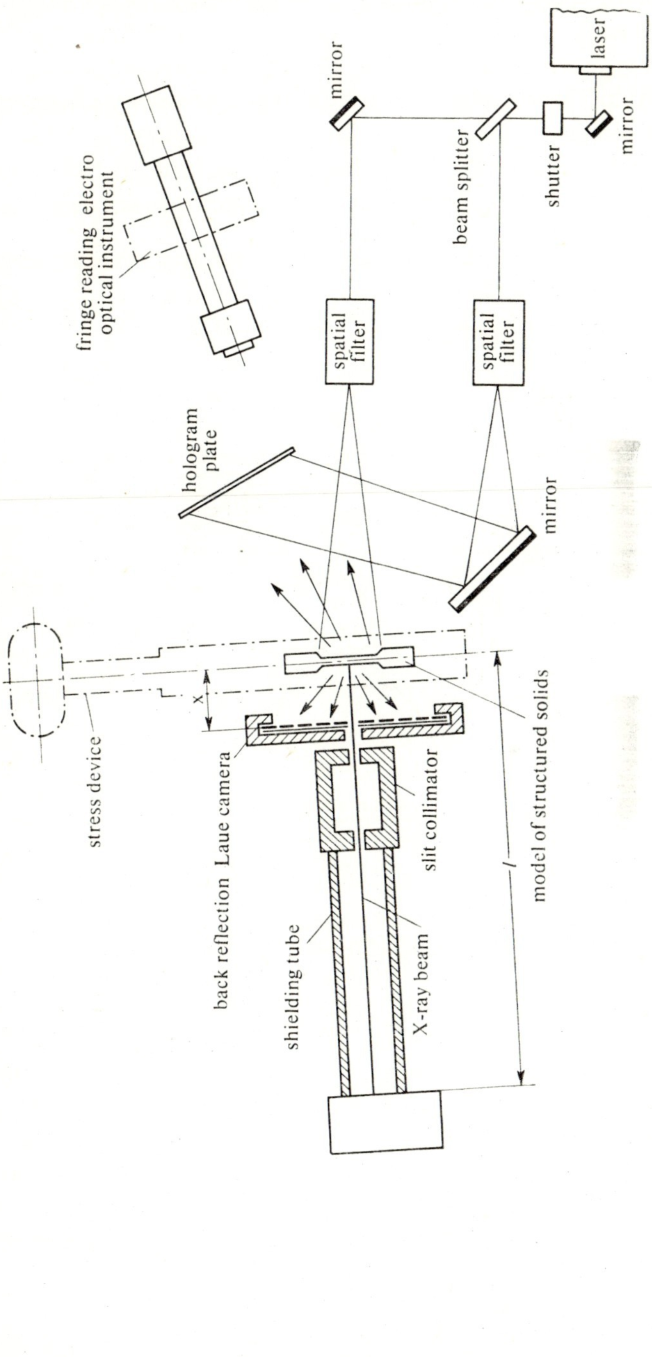

Fig. 42. Schematics of the X-ray and holographic interferometry set-up with the model of a crystalline solid (plan view).

the back-reflection Laue camera by using details in the stress application device not shown in the schematics. For the purpose of the X-ray diffraction technique the aluminium monocrystals had an "optically polished surface" that faced the Laue camera. For a more detailed discussion of the above technique the reader is referred to reference[134]. It should be noted, however, that in the performance of the X-ray tests, the exposure time of the film becomes significant.

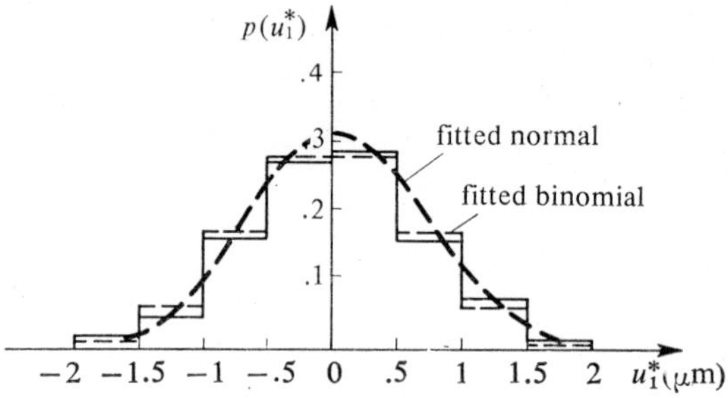

(a) Marginal probability density

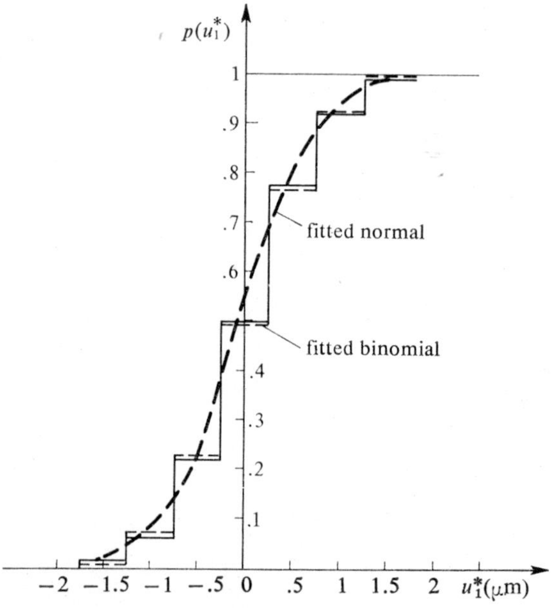

(b) Marginal probability distribution function

Fig. 43. Marginal distribution of the fluctuating part of microdeformations u_1.

Thus the latter depends largely on the distance l between the X-ray source and the material sample, the characteristics of the film itself, the type of screen in front of the film and finally on the intensity of the radiation. The achieved resolution and hence the accuracy of the tests also depend on l, the size of the slot h as well as the distance x of the specimen surface from the film (Fig. 42). It has been found in the performance of the above described tests, that good results

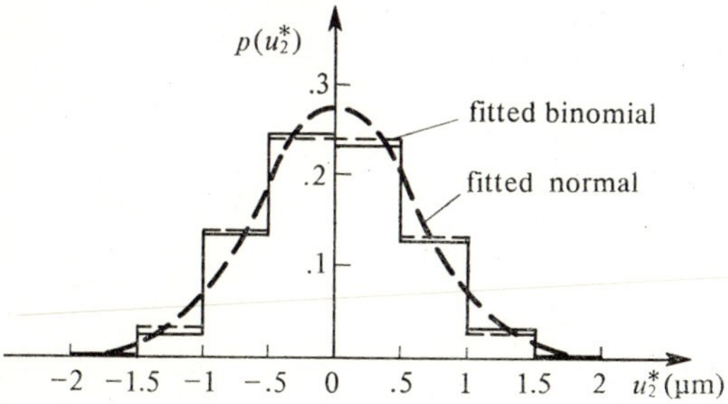

(a) Marginal probability density

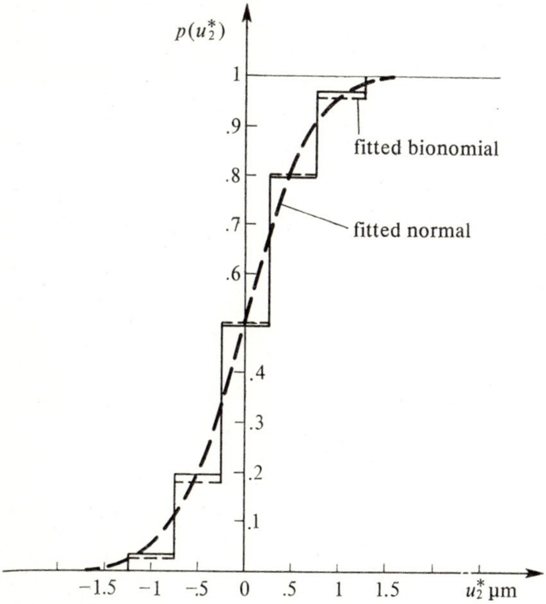

(b) Marginal probability distribution function

Fig. 44. Marginal distribution of the fluctuating part of microdeformation u_2.

can be obtained with an exposure time of two hours. Thus, the test procedure adopted for the combined holographic interferometry and X-ray diffraction technique has been as follows:

(i) First, an exposure of the X-ray film in the Laue camera for the undeformed material sample for two hours.

(ii) At the end of this period a hologram exposure of 10 seconds was carried out.

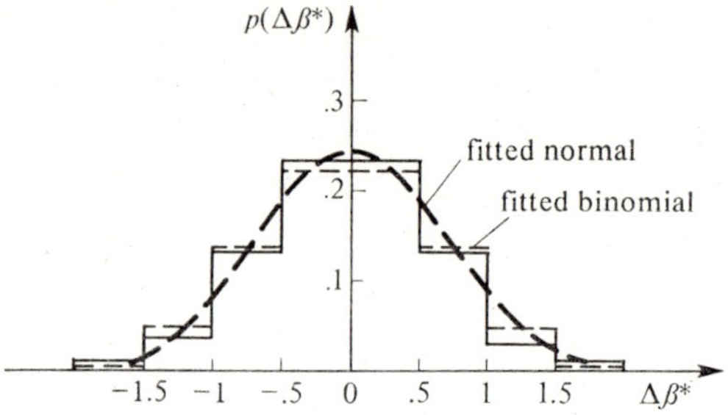

(a) Probability density

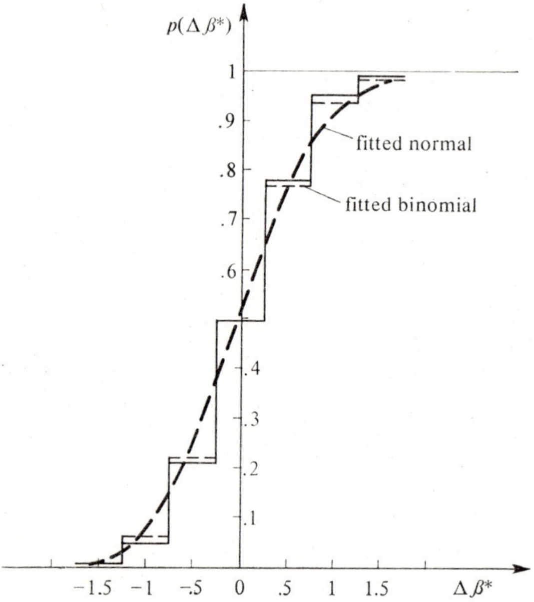

(b) Probability distribution function

Fig. 45. Distribution of microrotations $\Delta\beta^*$.

(iii) The material sample was then subjected to a stress level of 2.5 gm/mm^2 and 5 gm/mm^2, respectively (Figs. 34 and 35).

(iv) Immediately a second exposure of the hologram was performed followed by a second exposure of the X-ray film in the Laue camera.

The achieved results by using this combined technique are indicated in Figs. 43–45. Thus, Figs. 43 and 44 represent probability distributions of the microdeformation components u_1 and u_2 as well as their fluctuating parts u_1^* and u_2^* as discussed earlier. Fig. 45 shows the probability distribution of the microrotations of individual crystals in terms of the angle β in accordance with the geometry of the back-reflection technique (Fig. 38). All test values were obtained for the two-dimensional material model and the randomly distributed monocrystals as indicated in the arrangement of Figs. 40, 41. From the actual test, values were obtained in the form of histograms. The smooth curves shown were derived by using a normal fitting procedure. The diagrams show clearly the Gaussian character of the distributions as implicitly assumed in the general theory of stochastic deformations.

5.4 Experimental investigations of fibrous systems

Similar to the experimental work carried out on models of crystalline solids, the investigations concerning fibrous systems are also based on holographic interferometry in conjunction with other methods, in particular with that of scanning electron microscopy (SEM). However, in contrast to the work described in the foregoing section the studies of fibrous systems employed actual material samples such as "bond" and "newsprint" paper, that are essentially cellulosic systems. Although the experimental work described subsequently refers to this type of fibrous systems, the techniques discussed are equally applicable to other structures composed of synthetic fibres, polymers, etc.

Before dealing with the actual test arrangements, it is of interest to note, that the holographic interferometry yields results of three-dimensional surface deformations along various points of a "scanning line". These points however are of a finite size and represent a "scanning area" as mentioned in the model analysis of fibrous systems in Section 4.3 of Chapter IV. Measurements within certain scanning areas can only be performed by the aid of a scanning electron micro-

Fig. 46. Holographic set-up for testing fibrous systems.

scope technique and a relation between these two different observations must be brought about so that an overall response of the material sample to a specific load and environmental condition can be attained.

(A) *Holographic interferometry*

The holographic interferometry method applied to fibrous structures is somewhat different to that for crystalline solids since fibrous structures are very sensitive to environmental conditions such as humidity and temperature. As a consequence certain modifications in the stress device as well as in the other components of the instrumentation, i.e. optical set-up and fringe reading procedure are required. A general lay-out of the arrangement for the determination of the three-dimensional surface displacements of paper samples when subjected to uni-axial tensile stresses is given in the photograph of Fig. 46. This set-up has been used for the determination of the "thickness variation of thin paper strips" due to tensile loads which is of considerable interest in practical applications. It further indicates the optical components, the special stressing device used for testing fibrous structures including the environmental chamber, a

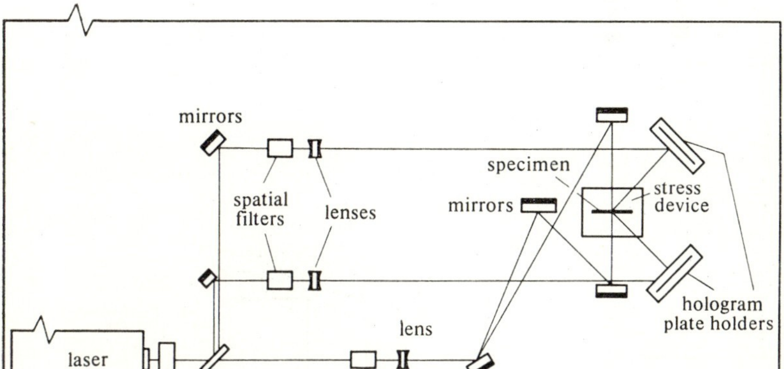

Fig. 47. Formation stage of holograms for two-sided observations on fibrous network.

special "scanning device" that is employed together with a photomultiplier to be discussed subsequently. In the stress apparatus a paper sample for instance can be stressed pneumatically between fixed and movable jaws within the environmental control chamber. The latter permitted the control of humidity between 35–98% within ±1% and a temperature control between 4–77°C with an accuracy of ±1/2°C.

The holographic interferometry technique as applied to fibrous systems to obtain three-dimensional displacement measurements again uses the single hologram exposure method, which in certain cases is extended to measure surface displacements on both sides of the material sample. This is indicated by Fig. 47, which shows also the shutter arrangement for the exact exposure timing, spatial filters, mirrors, beam splitters and expanding lens so that an equal intensity on both sides of the hologram plate can be obtained. As already mentioned, the method of fringe reading for crystalline solids (Section 5.3) is different from that in the present case, since the observed surface deformations for fibrous systems are of smaller magnitude and need therefore an even more accurate observation technique. Hence, the method employed here follows basically that described by Bellani and Sona[141], which employs a spherical mirror device as

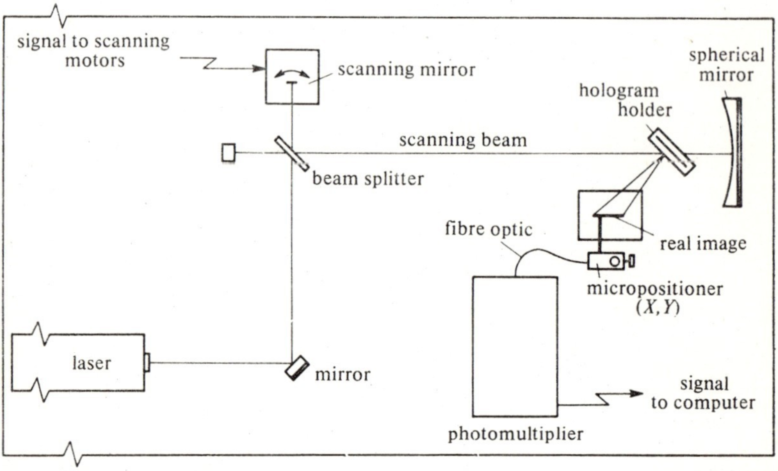

Fig. 48. Scanning method for automatic fringe readings on fibrous structures.

well as an automatically rotated "scanning mirror" that is driven by two scanning motors. This arrangement permits the fringe reading on the "real image" of the material sample to be carried out. The schematics of the set-up employed in these tests are given in Fig. 48, which also indicates the use of the "micro-positioner" for adjusting the "fibre optics" used as the sensor source that transmits the light absorption of the fringe pattern produced on the hologram as an input to the photo-multiplier device, the output of which is transferred to an analog-digital data computer. This method permits to read a large number of points and the corresponding dark or light fringes that move across them on each side of the material sample

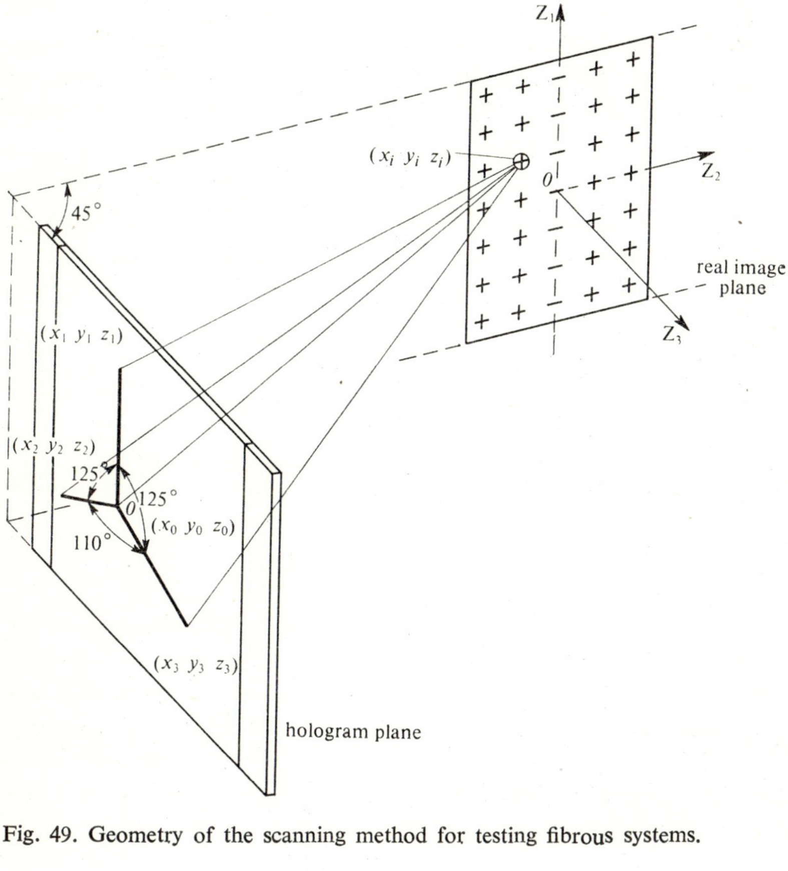

Fig. 49. Geometry of the scanning method for testing fibrous systems.

or more precisely on its real image. The evoluation of the obtained data computer output occurs in terms of equation (5.17) whereby the surface displacement vector ${}^{\alpha}\boldsymbol{u}$ at an object point is directly related to the number of fringes crossing this particular point. By continuously changing the line of observation or by using such a "scanning technique" it becomes necessary to adopt a different reconstruction geometry to that given in Fig. 33. Thus the scanning method requires a geometry as outlined schematically in Fig. 49 below. It is seen that three scanning directions can be chosen arbitrarily. However, to achieve good results and high accuracy of fringe readings (up to 1/8 of a fringe number), it has been found advantageous in the investigation of fibrous systems to use one vertical scanning direction and two left and right directions that include an angle of 125° with the vertical one. A more detailed discussion on this method is given in reference[126]. After obtaining the information about the fringe numbers from the data computer by using the geometry of Fig. 49, the following relations can be written for point "A" for instance:

$$\begin{bmatrix} l_1-l_0 & m_1-m_0 & n_1-n_0 \\ l_2-l_0 & m_2-m_0 & n_2-n_0 \\ l_3-l_0 & m_3-m_0 & n_3-n_0 \end{bmatrix} \begin{bmatrix} u_1 \\ u_2 \\ u_3 \end{bmatrix} = \pm \begin{bmatrix} \lambda N_1 \\ \lambda N_2 \\ \lambda N_3 \end{bmatrix} \tag{5.24}$$

in which N_1, N_2, N_3 are the fringe numbers and the values l_0, m_0, n_0 in the transformation matrix are obtained from:

$$\begin{aligned} l_0 &= \pm \frac{x_0-x_i}{[(x_0-x_i)^2+(y_0-y_i)^2+(z_0-z_i)^2]^{1/2}} \\ m_0 &= \pm \frac{y_0-y_i}{[(x_0-x_i)^2+(y_0-y_i)^2+(z_0-z_i)^2]^{1/2}} \\ n_0 &= \pm \frac{z_0-z_i}{[(x_0-x_i)^2+(y_0-y_i)^2+(z_0-z_i)^2]^{1/2}} \end{aligned} \tag{5.25}$$

and similarly for other points. A typical hologram obtained by the method indicated by Fig. 47 for a bond type paper structure subjected to a preload of 17 Kg and a load of 17.3 Kg by an incremental loading using the pneumatic stress device, environmental chamber and for an exposure time of one second of the film for each side of the material sample is shown in Fig. 50. The fringe pattern shows a distinct out-of-plane motion of the material sample, which occurs usually for thin strips of fibrous structures. The holographic interferometry technique briefly discussed in this paragraph is concerned with the observation and measurements of microdeformations of surface points ($i = 1, \ldots, 60$) of the material sample. It is important to note, however, that such points represent in reality a small area, which itself contains a number of fibres. Thus, the obtained deformation vectors acting at various surface points can only be associated with the geometrical centre of this area. The latter has been referred to earlier as a "scanning area". In order to bring about a correction between the fundamental quantities involved in the deformational behaviour of fibrous structures, e.g. the deformations measured by the above described technique and those occurring within the scanning area itself, it is evident that the holographic interferometry must be combined with scanning electron microscopy. It is only possible by such a combined method of observation to establish relations of the type indicated in Fig. 27 of Chapter IV. Thus, as shown previously, holographic interferometry leads to the establishment of the basic

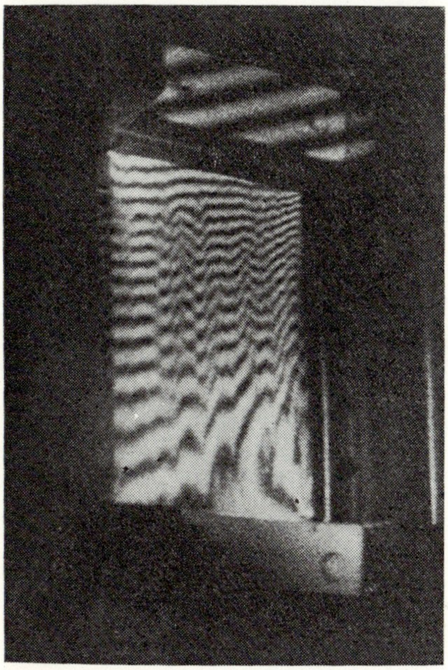

Fig. 50. Holographic-interferogram for bond paper at an incremental load of 300 gm.

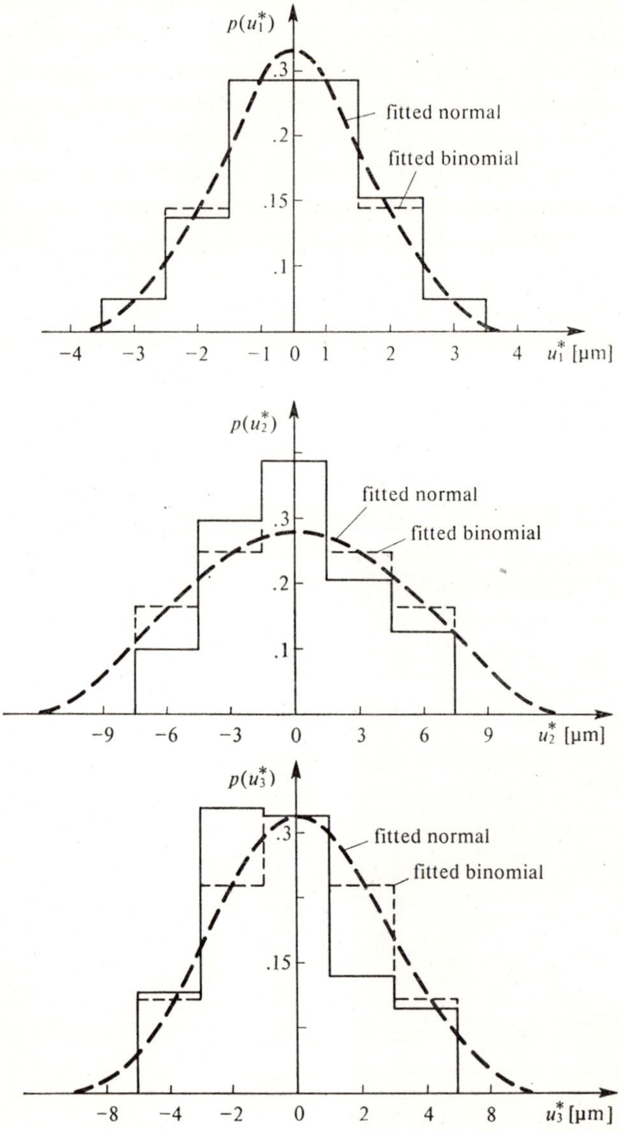

Fig. 51. Marginal distributions of the microdeformations (fluctuating parts) of "bond type paper" as obtained from holographic interferometry.

quantity ${}^{A}\boldsymbol{\Omega}$ (equation (4.95)). On the assumption that this quantity can be associated with the total deformation of an element (equation (4.99)), as obtained by scanning electron microscopy, then the inherent scale factor of these types of observations can be overcome. It is further of interest to note that for fibrous materials it is equally significant to find the profile of the "local basis weight" variation, which influences the overall response of such systems. The former is usually determined by means of β-radiography, which consists essentially of a radiation penetration technique (β-raygram) and which measures the mass per unit area of the sample as well as its distribution over the entire sample. It then becomes possible, by using an appropriate analysis, to correlate the local basis weight of the fibrous structure with the observed three-dimensional deformations caused by an applied load. Although in discussing the model analysis for cellulosic systems in Chapter IV, an analytical form for the basis weight has been given, no attempt at a correlation with the occurring deformations, as outlined in this paragraph, is made here. It may be noted that such an attempt and its discussion is given in reference[106]. Finally, the three-dimensional microdeformations for a "bond type paper" representing a fibrous system as obtained from holographic interferometry in terms of their corresponding distribution functions are given in Fig. 51. Apart from the use of scanning electron microscopy for the purpose mentioned above, its application is of utmost importance in the determination of the bonding behaviour between individual fibre elements. For this purpose a brief discussion on scanning electron microscopy and its application together with electron back-scattering and or cathodoluminescence is given below.

(B) *Scanning electron microscopy*

It has already been mentioned in Chapter IV, in connection with the analysis of fibrous systems, that one of the most significant parameters of the microstructure of such materials is the "relative displacement vector" arising from the motion of the otherwise "perfect bonding" arrangement between two structural elements. It has also been stated in the foregoing paragraph, that the scanning electron microscope technique becomes important in the establishment of a relation between the physical properties of a structural element or an ensemble of them, and the three-dimensional material sample of such a system. Scanning electron microscopy offers, however, a more versatile method of observation of surface characteristics of structural elements as those required above. Due to the limited scope of this text it is only possible to give a brief discussion on the possible use of scanning electron microscopy in conjunction with "cathodoluminescence" to gain some information as to the bonding behaviour between elements of a fibrous structure. In general, the essential features of a scanning electron microscope[142] can be stated as follows:

(a) it supplies an electron source,

(b) it permits focusing of electrons onto a small spot of the material specimen,

(c) it provides a possibility of scanning that spot across the material sample,

(d) it supplies means of detecting the response of the specimen by a display of the observations made,

(e) it contains also a transmission system for the response from the specimen to an appropriate display system.

The great advantage of a scanning electron microscope over an optical one, lies in its considerably higher resolution, its depth of focusing, its possibility of a transmission mode, reflection, diffraction, etc., a large enough field of view and finally the possibility to process a signal. Hence due to these properties a direct observation of an actual material sample at a high resolution can be made. It is further possible to switch over a wide range of magnifications so as to observe a rather fine detail of the material specimen which is at the same time readily identifiable as to its position within the entire material sample. It is important to note, however, that the electron range for scanning electron microscopy is the determining effect it has on the "spatial resolution" of a particular instrument. Thus, qualitatively speaking, for an electron beam to interact with a certain volume of the target material, the volume of the specimen to be sampled by the beam is in general smaller than the actual interaction volume and is defined by the limiting resolution of the instrument.

The scanning electron microscope employed at present for the study of bonding between natural and synthetic single fibres consists essentially of an electro-optical column with a completely pre-aligned four lens system. It contains the electron gun, the condenser lens system including the electron-optical column, liner tube and energizing coil in its lower part, and a final aperture. The instrument permits a magnification from 5 × up to 240000 ×, which is automatically linked to the accelerating voltage. The latter is variable between 1 KV to 30 KV with a stability of 10 ppm/10 min. The bias adjustment is also linked to the accelerating voltage. For the performance of bonding experiments the lower limit, i.e. 1–5 KV is at present employed. The scanning system of the instrument permits to record photomicrographs in the single frame recording mode, but also a visual mode for full frame observations with five speeds for careful detailed observations. It has a specially designed vacuum chamber, which is connected to the vacuum system which permits a normal operating vacuum of 2×10^{-5} to 2×10^{-6} torr, and for other requirements a step-up to 5×10^{-7} torr by the use of a liquid nitrogen trap and a LaB_6 filament. The vacuum chamber has ports for X-ray analysis experiments or other detecting devices as well as a port for transmission electron detector equipment. At present for the studies on bond strength between fibres a manually operated special "tensile stage" is used in conjunction with a special load-cell in the range of 0.5–500 gm and a video-amplifier system as well as two cathodoluminescence detectors. The latter use the property that some classes of materials emit light, when bombarded by high energy electrons. This phenomenon is well known as "cathodoluminescence". In the present case the specimen consisting of two bonded fibres relaxes under that bombardment back to its equilibrium configuration whilst emitting some of the absorbed energy from the primary electron beam as light. The luminescence decay time lies usually for the most efficient cathodoluminescent materials including organic crystals between 10^{-10} sec. to 100 sec. It is to be noted, however, that most luminescent materials are extremely sensitive to impurities and in many cases the luminescent spectrum is determined by species of impurity and the brightness of the concentration of the impurities. Hence, the method of cathodoluminescence can be used for the present purposes due to its sensitivity represented by only a few parts per million of impurity. The volume from which the luminescence emission comes is as large as the previously mentioned interaction volume of the primary electrons in the case of specimens that are transparent to their own cathodoluminescence emission. Hence the best resolution obtainable in these experiments will be comparable to that of the primary electron range.

In concluding this chapter it may be stated that even the limited experimental techniques discussed here, have shown that experimental information can be obtained for the significant parameters introduced in the general theory of stochastic deformations. Although in the past many experiments have been carried out with regard to single elements of a structured material or experiments that were concerned with the macroscopic response, no attempt has been made to develop techniques for closing the gap between these two types of observations.

Thus, it is hoped, that this brief study of experimental methods used in micromechanics and the indication of more sophisticated experimental procedures may lead ultimately to a more realistic representation of the overall response behaviour of structured solids within the framework of probabilistic micromechanics.

Bibliography

1 R. von Mises, *Mathematical Theory of Probability and Statistics*, Academic Press, New York, 1964.

2 A. N. Kolmogorov, *Foundations of Probability Theory*, Chelsea, New York, 1956.

3 M. Loeve, *Probability Theory*, Van Nostrand Co. Inc., Princeton, New Jersey, 1963.

4 B. V. Gnedenko, *Theory of Probability*, Chelsea, New York, 1963.

5 A. Rényi, *Foundations of Probability*, Holden-Day Inc., San Francisco, 1970.

6 Yu. V. Prohorov and Yu. N. Rozanov, *Probability Theory*, Springer-Verlag, New York, 1969.

7 V. S. Pugachev, *Theory of Random Functions*, Addison-Wesley Publishing Co., Mass., 1965.

8 A. M. Yaglom, *Theory of Stationary Random Functions*, Prentice-Hall, New York, 1962.

9 J. L. Doob, *Stochastic Processes*, John Wiley and Sons Inc., New York, 1953.

10 I. I. Gikhman and A. V. Skorohod, *Introduction to the Theory of Random Processes*, W. B. Saunders and Co., Philadelphia, 1969.

11 J. F. C. Kingman and S. J. Taylor, *Introduction to Measure and Probability*, Cambridge University Press, 1966.

12 K. R. Parthasarathy, *Probabilistic Measures on Metric Spaces*, Academic Press, New York, 1967.

13 L. Schwartz, *Theorie des Distributions*, Hermann, Paris, 1950.

14 I. N. Sneddon, in A. C. Eringen (ed.), *Continuum Physics I*, Academic Press, New York, 1971.

15 I. M. Gel'fand and G. E. Shilov, *Generalized Functions*, Vol. I, Academic Press, New York, 1964.

16 I. M. Gel'fand and N. Ya. Vilenkin, *Generalized Functions*, Vol. IV, Academic Press, New York, 1964.

17 I. M. Gel'fand, *Uspekhi Mat. Nauk*, 11 (1956) 3 (Russian).

18 J. Neveu, *Mathematical Foundations of the Calculus of Probability*, Holden-Day Inc., San Francisco, 1965.

19 D. A. Kappos, *Probability Algebras and Stochastic Spaces*, Academic Press, New York, 1969.

20 A. I. Khinchine, *Mathematical Foundations of Statistical Mechanics*, Dover Publ., New York, 1949.

21 G. D. Birkhoff, *Bull. Amer. Math. Soc.* (1932) 361.

22 R. Jancel, *Foundations of Classical and Quantum Statistical Mechanics*, Pergamon Press, London, 1963.

23 N. A. Friedman, *Introduction to Ergodic Theory*, Van Nostrand Reinhold Company, New York, 1970.

24 P. Billingsley, *Ergodic Theory and Theory of Information*, John Wiley and Sons Inc., New York, 1965.

25 R. V. Chacon and D. S. Ornstein, *Illinois J. Math*, 4 (1960) 153.

26 A. A. Markov, *Calculus of Probability*, (Russian) 4th Edition, Moscow, 1924.

27 A. N. Kolmogorov, *Bull. Math. Univ. Moscow*, 1 (1937) 16 (Russian)

28 W. Feller, *An Introduction to Probability and Its Applications*, Vol. I, John Wiley and Sons Inc., New York, 1957.

29 E. B. Dynkin, *Markov Processes*, Vols. I and II, Academic Press, New York, 1965.

30 A. T. Bharucha-Reid, *Elements of the Theory of Markov Processes and Its Applications*, McGraw-Hill, New York, 1960.

31 D. R. Axelrad and L. G. Jaeger, in M. Te'eni (ed.), *Proc. Southampton 1969 Civil Engineering Materials Conference*, Part I, Wiley Interscience, London, 1969, 571.

32 D. R. Axelrad and L. G. Jaeger, in M. Te'eni (ed.), *Proc. Southampton 1969 Civil Engineering Materials Conference*, Part I, Wiley Interscience, London, 1969, 87.

33 D. R. Axelrad, in S. Onagi (ed.), *Proc. 5th Int. Congress on Rheology*, 2, University of Tokyo Press, Tokyo, 1970, 221.

34 D. R. Axelrad and R. N. Yong, in S. Onagi (ed.), *Proc. 5th Int. Congress on Rheology*. 2, University of Tokyo Press, Tokyo, 1970, 309.

35 D. R. Axelrad and J. W. Provan, *Arch. Mech. Stos.*, 25 (1973) 811.

36 W. Bollmann, *Crystal Defects and Crystalline Interfaces*, Springer-Verlag, New York, 1970.

37 C. Goux *et al.*, *Surface Science*, 31 (1972) 115.

38 J. Yvon, *Correlation and Entropy in Classical Statistical Mechanics*, Pergamon Press, London, 1969.

39 P. and T. Ehrenfest, *The Conceptual Foundations of the Statistical Approach in Mechanics*, Cornell University Press, Ithaca, New York, 1959.

40 M. Kac, *Probability and Related Topics in Physical Sciences*, Lectures in Applied Mathematics, Vol. I, Interscience, New York, 1959.

41 G. E. Uhlenbeck, *The Statistical Mechanics of Non-Equilibrium Phenomenon*, Lecture Notes of Les Houches Summer School, 1955.

42 N. G. van Kampen, in E. D. G. Cohen (ed.), *Fundamental Problems in Statistical Mechanics*, North-Holland Publishing Co., Amsterdam, 1962.

43 D. R. Axelrad, *Arch. Mech. Stos.*, 23 (1971) 131.

44 D. R. Axelrad, *Random Theory of Deformation of Structured Media*, Lecture 71, Int. Cent. Mech. Sci., Udine, Italy, 1971.

45 D. R. Axelrad and J. W. Provan, *Thermodynamics of Deformation in Structured Media*, Lecture 71, Int. Cent. Mech. Sci., Udine, Italy, 1971.

46 D. R. Axelrad, *Introduction to the Probabilistic Micromechanics of Solids*, Institute for Mechanics Summer Course, University of Bochum, Federal Republic of Germany, 1974.

47 D. R. Axelrad, *Rheol. Acta.*, 12 (1973) 177.

48 G. L. Clark, *Applied X-Rays*, McGraw-Hill, New York, 1955.

49 B. D. Cullity, *Elements of X-Ray Diffraction*, Addison-Wesley Publishing Co., Mass., 1956.

50 W. H. Zachariasen, *Theory of X-Ray Diffraction in Crystals*, John Wiley and Sons Inc., New York, 1945.

51 R. W. James, *The Optical Principles of the Diffraction of X-Rays*, G. Bell and Sons Ltd., 1965.

52 D. R. Axelrad and J. Kalousek, in J. T. Pindera (ed.), *Mechanics in Research and Development*, Study No. 9, University of Waterloo, Canada, 1973.

53 G. Birkhoff, J. Bona and J. Kampé de Fériet, in A. T. Bharucha-Reid (ed.), *Probabilistic Methods in Applied Mathematics*, Vol. III, Academic Press, New York, 1968.

54 A. N. Kolmogorov and S. V. Fomin, *Elements of the Theory of Functions and Functional Analysis*, Vol. 2, Graylock Press, Albany, New York, 1961.

55 J. Dieudonné, *Foundations of Modern Analysis*, Academic Press, New York, 1960.

56 A. E. Taylor, *Introduction to Functional Analysis*, John Wiley and Sons Inc., New York, 1958.

57 R. C. Blumenthal and R. K. Getoor, *Markov Processes and Potential Theory*, Academic Press, New York, 1963.

58 G. F. Simmons, *Introduction to Topology and Modern Analysis*, McGraw-Hill, New York, 1963.

59 E. Hille and R. S. Philips, *Functional Analysis and Semi-Groups*, Amer. Math. Soc. Colloquium Publ., Vol. 31, 1957.

60 N. Bourbaki, *C.R. Acad. Sci., Paris*, 206 (1938) 1701.

61 C. Goffman and G. Pedrick, *First Course in Functional Analysis*, Prentice Hall Inc., New Jersey, 1965.

62 F. Treves, *Locally Convex Spaces and Partial Differential Equations*, Springer-Verlag, New York, 1967.

63 K. Yosida, *Functional Analysis*, Springer-Verlag, Berlin, 1965.

64 P. R. Halmos, *Measure Theory*, Van Nostrand Inc., Princeton, New Jersey, 1950.

65 S. Basu, *On a General Deformation Theory of Structured Solids*, Ph.D. Thesis, McGill University, Montreal, Canada, 1975.

66 N. Bourbaki, *Topologie Generale*, Chaps. III and IV, Hermann, Paris, 1951.

67 J. J. Moreau, *C.R. Acad. Sci., Ser. A*, 273 (1971) 118.

68 E. Tonti, *Quart. Res. Rep. Math., C.N.R.*, Italy, 1975.

69 S. Bochner, *Ann. Math.*, 48 (1942) 1014.

70 J. Kampé de Fériet, *Proc. Symp. Appl. Math., Hydrodynamic Instability*, 13 (1962) 165.

71 D. G. Kendall, *Trans. Am. Math. Soc.*, 78 (1955) 529.

72 A. Rényi, *Publ. Math.*, 2 (1951) 66.

73 A. Prekopa, *Acta. Math. Acad. Sci. Hungary*, 3 (1952) 317.

74 M. Fisz and K. Urbanik, *Studia Math.*, 15 (1956) 328.

75 M. Rosenblatt, *Markov Processes: Structure and Asymptotic Behaviour*, Springer-Verlag, New York, 1971.

76 D. R. Axelrad, J. W. Provan and S. Basu, in P. G. Glockner (ed.), *Proc. Conf. Symmetry, Similarity and Group Theoretic Methods in Mechanics*, University of Calgary, Canada, 1974.

77 V. I. Arnold and A. Avez, *Ergodic Problems in Classical Mechanics*, W. A. Benjamin Inc., New York, 1968.

78 S. R. Foguel, *The Ergodic Theory of Markov Processes*, Van Nostrand Reinhold Co., New York, 1969.

79 G. D. Birkhoff, *Proc. N.A.S.*, 17 (1931) 656.

80 S. Kakutani, *Proc. Imp. Acad. Tokyo*, 16 (1940) 49.

81 E. Hopf, *J. Rat. Mech. Anal.*, 3 (1954) 13.

82 D. R. Axelrad and S. Basu, in S. T. Ariaratnam, H. H. E. Leipholz, (eds.), *Stochastic Problems in Mechanics*, Study No. 10, University of Waterloo, Canada, 1973.

83 D. R. Axelrad and S. Basu, *Advances in Molecular Relaxation Processes*, 6 (1973) 185.

84 E. Tonti, Private Communications, 1975.

85 D. R. Axelrad, *Arch. Mech. Stos.*, 30 (1975).

86 M. S. Pinsker, *Information and Information Stability of Random Variables and Processes*, Holden-Day Inc., San Francisco, 1964.

87 W. A. Harrison, *Pseudo-potentials in the Theory of Metals*, W. A. Benjamin Inc., New York, 1966.

88 G. Leibfried, *Gittertheorie der Mechanischen und Thermischen Eigenschaften der Kristalle, Handbuch der Physik*, Band VII/1, Springer-Verlag, Berlin, 1955.

89 D. R. Axelrad, J. W. Provan and S. el Helbawi, *Arch. Mech. Stos.*, 25 (1973) 801.

90 D. R. Axelrad and J. W. Provan, *Symp. Stat. Cont. Mech.*, Polish Acad. Sci., Jabłonna, Poland, 1972.

91 A. H. Cottrell, *Theory of Crystal Dislocations*, Gordon and Breach, New York, 1964.

92 E. Kröner, *Kontinuumtheorie der Versetzungen und Eigenspannungen*, Springer-Verlag, Berlin, 1958.

93 E. Kröner, *Int. J. Engng. Sci.*, (1973).

94 J. P. Hirth and J. Lothe, *Theory of Dislocations*, McGraw-Hill, New York, 1968.

95 H. Zorski, *Int. J. Solids Struct.*, 4 (1968) 959.

96 A. Seeger, *Moderne Probleme der Metallphysik*, Vols. I and II, Springer-Verlag, Berlin, 1965.

97 D. C. Wallace, *Thermodynamics of Crystals*, John Wiley and Sons Inc., New York, 1972.

98 F. C. Frank, *Phys. Rev.*, 79 (1951) 722.

99 J. W. Christian, *The Theory of Transformations in Metals and Alloys*, Pergamon Press, Oxford, 1965.

100 A. V. Granato, *Internal Friction Studies of Dislocation Motion*, Dislocation Dynamics, McGraw-Hill, New York, 1968.

101 J. W. Provan and D. R. Axelrad, *Arch. Mech. Stos.*, 30 (1975).

102 J. W. Provan and O. A. Bamiro, *Canadian Metal. Q.*, 13 (1974) 1.

103 O. A. Bamiro, *On Elastic Grain Boundary Effects in Polycrystalline Solids*, Ph. D. Thesis, McGill University, Montreal, Canada, 1975.

104 Yu. N. Rabotnov, *Creep Problems in Structural Members*, Wiley Interscience, New York, 1969.

105 N. Distefano and R. Todeschini, *Int. J. Solids, Struct.*, 9 (1973) 805.

106 Y. M. Haddad, *Response Behaviour of a Two-Dimensional Fibrous Network*, Ph. D. Thesis, McGill University, Montreal, Canada, 1975.

107 L. Pauling, *The Nature of the Chemical Bond and the Structure of Molecules and Crystals*, Cornell University Press, Ithaca, New York, 1960.

108 D. Hadzi and W. H. Thompson, *Hydrogen Bonding*, Pergamon Press London, 1959.

109 P. M. Morse, *Phys. Rev.*, 34 (1929) 57.

110 N. D. Sokolov, in D. Hadzi and W. H. Thompson (eds.), *Symp. Hydrogen Bonding*, Pergamon Press, London, 1959.

111 H. Corte and O. J. Kallmes, in F. Bolam (ed.), *Trans. Oxford Symp.* Tech. Sec., Paper and Board Makers' Assoc., London, 1962.

112 V. Ya. Andreichenko, *Sb. Tr. VNII Tsellyul. Bumazh. Prom.*, 58 (1971) 80.

113 O. J. Kallmes, *Tappi*, 52 (3) (1969).

114 D. R. Axelrad, D. Atack and J. W. Provan, *Rheol. Acta.*, 12 (1973) 170.

115 D. R. Axelrad, Y. M. Haddad and D. Atack, in G. J. Dvorak (ed.), *Proc. 11th Annual Meeting, Soc. Engng. Sci.*, Duke University, North Carolina, 1975.

116 D. R. Axelrad, S. Basu and D. Atack, *Proc. 7th Int. Congress on Rheology*, Gothenburg, Sweden, 1976.

117 D. R. Axelrad, *Advances in Molecular Relaxation Processes*, 2 (1970) 41.

118 D. R. Axelrad and J. W. Provan, *Rheol. Acta.*, 10 (1971) 330.

119 D. R. Axelrad, *Proc. 7th Int. Congress on Rheology*, Gothenburg, Sweden, 1976.

120 D. R. Axelrad, *14th Int. Congress of Theoretical and Applied Mechanics*, IUTAM Delft, The Netherlands, 1976.

121 E. Tonti, *Meccanica*, 5 (1970) 22.

122 E. F. Kaelble, *Handbook of X-Rays*, McGraw-Hill, New York, 1967.

123 P. Grivet, *Electron Optics*, Pergamon Press, Oxford, 1972.

124 L. E. Murr, *Electron Optical Applications in Material Science*, McGraw-Hill, New York, 1970.

125 R. D. Heidenrich, *Fundamentals of Transmission Electron Microscopy*, Interscience, New York, 1964.

126 D. R. Axelrad, S. Basu and Ri. Peralta-Fabi, *Experimental Investigation of Three-Dimensional Changes under Tensile Loading of Thin Fibrous Structures*, *Experimental Mechanics* (to be published).

127 D. R. Axelrad and J. Kalousek, *Micromechanics Lab. Rep. 71-7*, McGill University, Montreal, Canada, 1971.

128 R. J. Collier, C. B. Burckhardt and L. H. Lin, *Optical Holography*. Academic Press, New York, 1971.

129 R. K. Erf, *Holographic Nondestructive Testing*, Academic Press, New York, 1974.

130 J. Goodman, *Introduction to Fourier Optics*, McGraw-Hill, New York, 1968.

131 G. W. Stroke and A. E. Labeyrie, *Phys. Letters.*, 20 (4) (1968) 368.

132 E. B. Alexandrov and A. M. Bonch-Bruevich, *Sov. Phys., Tech. Phys.*, 12 (2) (1967) 258.

133 A. E. Ennos, *J. Sci., Instr.*, 2 (1) (1968) 731.

134 J. Kalousek, *Experimental Investigations of the Deformation of Structured Media*, Ph. D. Thesis, McGill University, Montreal, Canada, 1973.

135 P. M. Sutton, *Phys. Rev.*, 91 (1953) 816.

136 K. Lücke and S. Bühler, *Proc. Conf. on The Relation between the Structure and Mechanical Properties of Metals*, National Phys. Lab., London, 1963,

137 P. F. Kane and G. R. Larrabcc, *Characterization of Solid Surfaces*, Plenum Press, New York, 1974.

138 W. L. Bragg, *The Crystalline State*, McMillan and Co., New York, 1934.

139 E. E. Underwood, *Quantitative Stereology*, Addison-Wesley Publishing Co., 1970.

140 C. S. Barrett, *Structure of Metals*, McGraw-Hill, New York, 1952.

141 V. F. Bellani and A. Sona, *Applied Optics*, 13 (6) (1974) 1337.

142 J. W. S. Hearle, J. T. Sparrow and P. M. Cross, *The Use of the Scanning Electron Microscope*, Pergamon Press, Oxford, 1972.

Snbject index